ADVANCES IN
GROUP PROCESSES

Volume 12 • 1995

ADVANCES IN GROUP PROCESSES

Volume Editors: **BARRY MARKOVSKY**
KAREN HEIMER
JODI O'BRIEN
Department of Sociology
University of Iowa

Series Editor: **EDWARD J. LAWLER**
Department of Organizational Behavior
Cornell University

VOLUME 12 • 1995

 **JAI PRESS INC.**

Greenwich, Connecticut   *London, England*

Copyright © 1995 *JAI PRESS INC.*
55 Old Post Road, No. 2
Greenwich, Connecticut 06836

JAI PRESS LTD.
The Courtyard
28 High Street
Hampton Hill
Middlesex TW12 1PD
England

ISBN: 1-55938-872-2

Manufactured in the United States of America

CONTENTS

LIST OF CONTRIBUTORS

William Sims Bainbridge
Sociology Program Director
National Science Foundation

Mouraine R. Baker
Department of Sociology
University of Iowa

Karen A. Bantel
School of Business Administration
Wayne State University

Peter L. Callero
Department of Sociology
Western Oregon State College

Mark Chaffee
Department of Sociology
University of Iowa

Sydney Finkelstein
Tuck School of Business
Dartmouth College

Karen Heimer
Department of Sociology
University of Iowa

Cathryn Johnson
Department of Sociology
Emory University

Joseph Kasof
Department of Sociology
University of Texas at Austin

Theodore D. Kemper
Department of Sociology
St. John's University

Michael D. Large
Department of Sociology
University of Iowa

Edward J. Lawler
Department of Organizational
  Behavior
Cornell University

Barry Markovsky
Department of Sociology
University of Iowa

Jodi O'Brien   Department of Sociology
                                        University of Iowa

Yuriko Saito   Department of Sociology
                                        Kwansei Gakuin University

Robin W. Simon   Department of Sociology
                                        University of Iowa

Jeongkoo Yoon   Department of Organizational
                                            Behavior
                                        Cornell University

PREFACE

EDITORIAL POLICY

The purpose of this series is to publish theoretical analyses, reviews, and theory-based empirical chapters on group phenomena. We adopt a broad conception of "group processes." This includes work on groups ranging from the very small to the very large, and on classic and contemporary topics such as status, power, exchange, justice, influence, decision making, intergroup relations, and social networks. Contributors to the series have included scholars from diverse fields including sociology, psychology, political science, and organizational behavior.

The series provides an outlet for papers that may be longer, more theoretical and/or more integrative than those published by standard journals. We place a premium on the development of testable theories and on theory-driven research programs. Chapters in the following categories are especially apropos:

- *Conventional and unconventional theoretical work, from broad meta-theoretical and conceptual analyses to refinements of existing theories and hypotheses.* One goal of the series is to advance the field of group processes by promoting theoretical work.
- *Papers that review and integrate programs of research.* The current structure of the field often leads to the piecemeal publication of different parts of a program of research. This series offers those engaged in programmatic research on a given topic an opportunity to integrate their

published and unpublished work into a single paper. Review articles that transcend the author's own work are also of considerable interest.

• *Papers that develop and apply social psychological theories and research to macrosociological processes.* One premise underlying this series is that links between macro and microsociological processes warrant more systematic and testable theorizing. The series encourages development of the macrosociological implications embedded in social psychological work on groups.

In addition, the editors are open to submissions that depart from these guidelines.

CONTENTS OF VOLUME 12

The ten chapters in this volume cover a broad range of substantive topics and theoretical approaches. Three chapters address issues pertaining to the consequences of individual attraction to social groups. Sociologists of religion debate whether the number and diversity of available religious denominations have positive or negative effects on the proportion of a population affiliated with religious groups. Bainbridge employs the Stark-Bainbridge theory of religion to address this problem. The theory offers basic axioms pertaining to, among other things, reward-seeking, social exchange, and environmental contingencies as these factors influence the religious behavior of individuals and collectivities. Markovsky and Chaffee extend the Markovsky-Lawler theory of group solidarity (from Volume 11) that provided a network-based conceptualization of group structure. New to this extension is the inclusion of social identification as a potential relational basis for group solidarity, and a conceptualization of some of the dimensions along which disintegrative forces may vary. Lawler, Yoon, Baker, and Large propose that token gifts in ongoing exchange relations serve expressive functions, and consequently foster commitment and closeness in those relations. They report on experimental tests providing support for their theory.

A number of the chapters forge new linkages between micro and macrosociological phenomena. Simon uses a role-theoretical perspective to examine gender differences in reference groups and self-evaluations among married, employed persons. The findings suggest that combining work and family roles yields a greater mental health benefit for men than for women. Kemper proposes that five constructs—division of labor, social relations of power and status, culture, composition, and ecology—together provide a comprehensive framework for studying individual outcomes in social contexts. In addition, he reports that in a large sample of recent journal publications, sociological variables are virtually always explained by reference to one or more of these constructs.

Several other chapters also bridge traditional micro and macro boundaries. Saito describes a new distributive justice theory. It builds on an earlier theory, highly formalized and empirically grounded, that explains how members of a society form mental images of stratified social class systems. Saito extends the earlier theory by using comparisons among components of these class system images as determinants of judgments of a society's fairness. Callero has developed a role-as-resource formulation and, in his chapter, applies it in an analysis of an uprising in a women's prison. He examines how the role of "woman inmate" is used as a source of agency to attempt to alter the social structure of the prison. Bantel and Finkelstein develop a model of management teams that links decision making to group demographics. Certain attributes of top team structures are assumed to provide special insights that enhance strategic decisions and organizational performance. The authors then show how these demographic characteristics are the outcomes of organizational, environmental, and strategic factors.

Two chapters apply the theory of status characteristics and expectation states. Johnson examines sexual orientation as a diffuse status characteristic, capable of invoking a variety of differentially evaluated inferences on the part of social interactants. She argues that lesbian/gay orientations are devalued compared to straight sexual orientations, and that this has important implications for interactions in task groups. Kasof's application demonstrates how expectations based on relative status may influence the way original products are received by group members, and how creativity thus comes to be socially and situationally defined. This approach departs in critical ways from prior work that focused on the dispositions and capacities of isolated individuals.

Barry Markovsky
Karen Heimer
Jodi O'Brien

Volume Co-editors

SOCIAL INFLUENCE AND RELIGIOUS PLURALISM

William Sims Bainbridge

ABSTRACT

Sociologists of religion currently debate whether the proportion of the population affilated with religious groups is increased or decreased by the diversity of denominations available in the community. Both viewpoints can be rooted in a general theory of religion that sees faith as the result of individual needs and processes of social exchange. Probably, both denominational diversity and denominational monopoly have effects that can increase religious mobilization, dependent upon the context of other factors, and the balance of these opposing forces will shift from one data set to the next, and from one set of control variables to another. Rather than causing sociologists to despair at the possibility of conducting conclusive research, this complex situation should inspire us to undertake fresh research projects directly examining the social and cultural dynamics of religious groups.

Contemporary theories and much recent research in the sociology of religion are strongly oriented toward group processes, social networks, and

Advances in Group Processes, Volume 12, pages 1-18.
ISBN: 1-55938-872-2

interpersonal exchange. However, the considerable accomplishments in this subdiscipline have as yet had little impact on sociology and social psychology generally. Furthermore, researchers on group processes who lack backgrounds in the sociology of religion seem unaware of the opportunities to test or develop general theories via empirical studies of churches, religious movements, and the eminently interactional processes of recruitment and defection. Thus an analysis of a major controversy in the sociology of religion may not be out of place in a volume on group processes.

A current debate centers on two opposing arguments about the relationship between religious pluralism and commitment (Warner 1993). In a series of publications, Roger Finke and Rodney Stark (1988, 1989a, 1989b, 1992; Finke, 1989) have pictured religion as a market economy in which denominations compete with each other for members. Different individuals and groups in society have different needs, cultures, and non-religious affiliations, so therefore religious pluralism should increase commitment by offering each person the style of religion that suits him or her best. In their empirical work, Finke and Stark have tried to show that rates of church membership are higher where there are more denominations in the religious marketplace.

In contrast, other researchers have argued that religious pluralism has a negative effect on church membership (Breault 1989a, 1989b; Land, Deane, and Blau 1991; Blau, Land, and Redding 1992; Blau, Redding, and Land 1993; cf. Christiano 1987). Religious monopoly might be associated with higher rates of religious involvement, if individual affiliations are chiefly the result of social influence, and if social influence is most effective when it is monolithic.

Thus, the narrow debate over denominational diversity and religious mobilization bears directly upon two distinctive general models of group process. The diversity-mobilization argument conceptualizes group affiliation in terms of individual choices among competing suppliers, with individuals maximizing their satisfaction by selecting the suppliers that best meet their personal needs. The monopoly-mobilization argument sees affiliation in terms of the net power of social influences operating within a diffuse social network, wherein persons are more strongly impelled to join a group the greater the proportion of their consociates who are members.

In the vigorous debate over proper methods of empirical research that has raged in the journals over this issue, the theoretical underpinnings have tended to become obscured. It is widely assumed that the diversity-mobilization hypothesis is rooted in the Stark-Bainbridge theory of religion (Stark and Bainbridge 1985, 1987; cf. Finke and Stark 1988), which laid the groundwork for a market analysis of religion and sought to establish the sociology of religion on a more rigorous footing (Simpson 1990; Collins 1993). However, in truth the monopoly-mobilization hypothesis also can be derived from the Stark-Bainbridge theory, and a close examination of how both of the competing hypotheses blend into a unified theory of group processes will reveal much about the limitations and prospects of contemporary sociology.

A THEORY OF RELIGION

The Stark-Bainbridge theory seeks to derive a long series of formal propositions about human exchange and religion, from a very short list of "axioms," by means of logical deduction and a set of strict definitions. Admittedly, the arguments it provides are only the outlines of possible proofs rather than fully stated rigorous deductions, and given the scope of the theory considerable effort would be required to fill all the gaps. In general, the theory accepts the ideas of George C. Homans (1974, 1984) about what a theory should do and how it should be constructed. A formal theory is a system of propositions, that logically derives relatively specific statements from relatively general ones, thereby explaining empirical observations in terms of general laws.

The theoretical system consists of a logical structure connecting three kinds of statements: axioms, definitions, and propositions. There are seven axioms, all to be quoted subsequently, comparable to the traditional axioms of plane geometry. Definitions associate each term to be defined with a particular set of properties or conditions that must be satisfied for something to be declared to be an instance of that term. Propositions, equivalent to the theorems of plane geometry, are relatively general statements that can be derived, directly or indirectly, from axioms and definitions. For convenience, the appendix of *A Theory of Religion* lists the axioms, numbered A1 through A7, the most important definitions, Def. 1 through Def. 104, and a swarm of propositions, P1 through P344. Many of the propositions do not concern religion, per se, but are statements about human interaction that were needed as stepping stones to propositions that were narrowly focused on religion.

The theory begins with a statement so obvious that social scientists hardly ever notice its importance: "A1 Human perception and action take place through time, from the past into the future." Behaviorism and its modern sociological equivalent, social learning theory, assert that human behavior is conditioned by past events. Rational choice theory argues that behavior is oriented toward expected future consequences of actions. There is no disagreement here, only a difference of emphasis.

Past experiences, especially the contingencies of prior reinforcement, give people conceptual frameworks for anticipating the value of future events and selecting among actions that will influence the future. "A2 Humans seek what they perceive to be rewards and avoid what they perceive to be costs." Although some sensations are directly rewarding, most human experience concerns things and events that are merely stages on the way to palpable rewards, what are sometimes called instrumental rewards. Thus, much of the time, humans seek things and situations that they believe will be rewarding, but they can never be sure.

Beyond the distinction between palpable (or primary) rewards and instrumental (or secondary... tertiary... etc.) rewards, many other categories and dimensions could be identified. "A3 Rewards vary in kind, value, and generality." Because rewards vary in kind, humans can profit from exchanges with each other, and there is no one best way of obtaining all rewards. Among the instrumental rewards are various pieces of information that guide us in obtaining other rewards. Rewards are general to the extent that they include other (less general) rewards.

Given the variety of rewards that may be sought, and our good fortune to be scions of a rather successful history of biological evolution, humans are mentally equipped to follow complex plans for the attainment of various goals. "A4 Human action is directed by a complex but finite information-processing system that functions to identify problems and attempt solutions to them." This system, which consists of the "hardware" of the brain as well as the "software" of culture, is commonly called the mind.

At this point in the development of the theory, it became possible to sketch the derivation of propositions. Especially important are propositions that introduce the concept of explanation: "Def. 10 *Explanations* are statements about how and why rewards may be obtained and costs are incurred." Thus, explanations are instructions, recipes, plans, indeed theories but ones with practical significance. Among the many strange qualities of sociology, perhaps the least defensible is the discipline's general failure to pay attention to skills and practical knowledge. The "sociology of knowledge" is really the sociology of political ideology, as if there were nothing worth knowing in this life but the "fact" that the bourgeoisie exploits the proletariat. At root, the human mind is neither political nor philosophical but practical. "P3 In solving problems, the human mind must seek explanations." "P4 Explanations are rewards of some level of generality."

Thus, the theory of religion is basically utilitarian, but it is also tragic. "A5 Some desired rewards are limited in supply, including some that simply do not exist." "A6 Most rewards sought by humans are destroyed when they are used." "A7 Individual and social attributes which determine power are unequally distributed among persons and groups in any society." These three axioms mean that some pressing human desires will always be frustrated, that satisfaction quickly turns to dissatisfaction, and that some people are even worse off than others in what is inescapably a disappointing world.

The ray of sunshine in this gloom is social exchange. "P6 In pursuit of desired rewards, humans will exchange rewards with other humans." Because rewards are consumed, people will seek the same one repeatedly, thus placing a high priority on gaining a good explanation on how to do so reliably and efficiently. The wide variation in the distribution of rewards means that some people are ready and willing suppliers of a particular reward, accepting a different reward in exchange for it through recurring interactions with the same other individuals. Thus arise relationships and social networks.

Explanations are among the rewards most commonly obtained through social exchange. Unfortunately, but consistent with the tragedy of the human condition, explanations on how to obtain scarce and nonexistent but intensely desired rewards are often very difficult to test empirically. Therefore, they spread relatively unchecked through networks of communication. "P14 In the absence of a desired reward, explanations often will be accepted which posit attainment of the reward in the distant future or in some other non-verifiable context." "Def. 18 *Compensators* are postulations of reward according to explanations that are not readily susceptible to unambiguous evaluation."

The greatest unmet needs, such as everlasting life with good health and unlimited love, bring people to religion. "P22 The most general compensators can be supported only by supernatural explanations." "Def. 20 Compensators which substitute for a cluster of many rewards and for rewards of great scope are called *general compensators*." "Def. 22 *Religion* refers to systems of general compensators based on supernatural assumptions."

THE MONOPOLY-MOBILIZATION THEORY

The idea that consensus strengthens religion arises in a discussion of the historical emergence and social role of religious specialists (Stark and Bainbridge 1987, pp. 89-96). Why should people give material and social rewards to priests who promise to mediate on their behalf with the gods? Why do they not simply talk to the gods directly? The answers are somewhat involved, but a few key points can be stated briefly.

People seldom have the opportunity to evaluate fully an explanation they receive through social exchange. Yes, if action based on it manifestly delivers the desired reward, it is a good explanation. But the person does not know whether it is the best one, because some other set of actions might deliver the same reward more cheaply. The collective experience of many people often provides a better judgment than the limited knowledge of just one person, especially with rewards that are seldom unambiguously obtained through simple actions. Thus, even in the most practical realms of human life, people often rely upon others to evaluate explanations for them. This is especially so when the explanations are very difficult to test through personal experience, as is the case with religious compensators.

In their interaction with other individuals, people learn which exchange partners to trust, and they will develop mental lists of the people to consult about particular problems. An individual's exchange partners will have similar lists, and by communicating about their problems people develop a rough consensus concerning whom to exchange with for a given reward. Thus, professional specialists will arise in the historical development of human culture, as soon as the particular field of endeavor had developed a plausible subculture. Because it was not constrained by the harsh empirical failures that could discredit professions rooted in physical technologies, religion was one of the very first professions to arise. In a sense it bootstrapped itself into existence, as priests conspired to convince the laity that their particular brand of religion was the only valuable one, often in a monopolizing alliance with the state.

A string of theoretical propositions outlines the argument: "P13 The more valued or general a reward, the more difficult will be evaluation of explanations about how to obtain it." "P62 No human being can personally evaluate all the explanations he uses, including verifiable ones." "P63 The value an individual places on an explanation is often set by the values placed on it by others and communicated to him through exchanges." "P64 In the absence of a more compelling standard, the value an individual places on a reward is set by the market value of that reward established through exchanges by other persons." "P65 The value an individual places on a general compensator is set through exchanges with other persons." "P66 When there is disagreement over the value of an explanation, the individual will tend to set a value that is a direct averaging function of the values set by others and communicated to him through exchanges, weighted by the value placed on such exchanges with each partner."

This chain of very general observations suddenly focuses again on religion. "P67 The more cosmopolitan a society with respect to religious culture, the lower the market value of any given general compensator." "Def. 36 *Cosmopolitan* refers to the existence of plural cultures within a society." Competing denominations are plural cultures, and thus diversity is equivalent to religious cosmopolitanism. To the extent that denominations differ in doctrine or practice, a person socially connected to two or more will experience lowered faith, compared to someone living in a religiously monolithic society where there are no religious disagreements.

THE DIVERSITY-MOBILIZATION THEORY

The pluralism argument arises from a discussion of the origins of sects (Stark and Bainbridge 1987, pp. 141-149). The hallmark of religion is that it provides general compensators that substitute for the set of major rewards, such as everlasting life, of which all human beings are deprived. Thus, one would think that a single Church Universal could provide the same comfort to everyone.

But this ignores the fact that some highly desired rewards are actually possessed by some persons while denied to others. Health and wealth are good examples. The wealthy may tend to be healthy, compared with the very poor, but ill health is found in all social groups. In contrast, poverty and wealth tend to be concentrated in very different corners of a community's social network. This means that deep social cleavages may arise separating the rich from the poor, whereas the healthy and unhealthy are found in every group and most families. Thus it is not surprising that compensators associated with social stratification have often been the basis for schisms that turn a unified church into two daughter organizations, a church and a sect.

For example, the members of a church congregation are often people with relatively high social status in their secular communities, whereas the members of a sect congregation are of lower secular status. To some extent, the church sanctifies the existing social status of its members. The sect, in contrast, provides compensatory status. To the sect member, possession of wealth and high education are not badges of honor; rather, faith in God and membership in the sect are the most noble qualities, conferring subjective status. In the extreme, the sect will define as sinful the consumption habits of members of the church congregation, while the church defines as presumptuous the status claims of sect members.

In terms of the theory's definitions and propositions, the argument proceeds according to the following outline: "Def. 57 A *sect movement* is a deviant religious organization with traditional beliefs and practices." "Def. 59 *Deviance* is departure from the norms of a culture in such a way as to incur the imposition of extraordinary costs from those who maintain the culture." "Def. 60 A *schism* is the division of the social structure of an organization into two or more independent parts." "P140 Consumers of scarce rewards and consumers of the corresponding compensators tend to avoid relationships with each other." "P142 Tension with the surrounding sociocultural environment is equivalent to subcultural deviance, marked by difference, antagonism, and separation." "P143 To the extent that religious groups are involved with compensators for scarce rewards, they are in tension with the sociocultural environment." "P185 The more socially closed a group, the more readily it can generate and sustain compensators, through exchanges among members."

The underlying logic of this argument is that monolithic religious denominations cannot simultaneously meet the contrasting needs of people who possess highly desired scarce rewards and people who lack them. Given a modicum of religious freedom, the result is schism and the eruption of high-tension sect movements. Because these tend to be socially encapsulated and begin life with relatively small memberships, a society will tend to have many of them. But once several sects exist, they will collectively serve a substantial fraction of the relatively deprived classes in the society. The denominations they broke out of will be even better positioned to serve the needs of relatively advantaged persons, because they will feel fewer demands to serve the disadvantaged. The result of sect formation, therefore, will be increased mobilization: "P153 In a culture with a dominant religious tradition, the emergence of sect movements increases the proportion of the population affiliated with religious organizations."

The conditional with which this proposition begins is important in two ways. First, if the society has a dominant religious tradition, shared by denominations and sects alike, then the doctrines and practices of low-tension and high-tension groups will differ only in degree, not kind. That is, both liberal Protestant denominations and radical Protestant sects agree on the divinity of Jesus, the efficacy of prayer, and many other important points. Competing Protestant groups do not contradict each other completely, but only in circumscribed areas such as the proper attributes of social status. A Protestant society, therefore, has many of the advantages of religious monopoly despite possessing religious diversity.

Second, if the society lacks a dominant religious tradition, then religious movements are much more free to undertake radical religious innovation and become cults. "Def. 58 A *cult movement* is a deviant religious organization with novel beliefs and practices." "P219 Persons who desire limited rewards that exist, but who lack the social power to obtain them, will tend to affiliate with sects, to the extent that their society possesses a dominant religious tradition supported by the elite." "P220 Persons who desire limited rewards that exist, but who lack the social power to obtain them, will tend to affiliate with cults, to the extent that their society does not possess a dominant religious tradition supported by the elite." Religious innovation may lay the groundwork for entirely new religious traditions that eventually gain substantial power in society, as was the case with Christianity in the Roman Empire and Mormonism today.

Healing cults, especially, can serve needs for relatively specific compensators that cut across social classes. Thus, a society like contemporary America that possesses many sects and cults would appear to have a highly fruitful form of diversity that by serving many different patterns of need should achieve a high level of religious mobilization. However, the memberships of cults remain quite small, and cults seldom appear in the measures of diversity that have been employed by published empirical studies concerning the mobilization theories. And one would think that very active cultic subcultures would undercut the plausibility of any particular set of compensators, because they disagree with respect to the most general compensators as well as specific ones, thereby greatly reducing mobilization. So there may be good reason to limit the diversity-mobilization theory to sects and denominations within a dominant tradition, as the original Stark-Bainbridge theory does.

THE EQUILIBRIUM PROBLEM

The fact that the Stark-Bainbridge theory leads to apparently contradictory predictions is not the flaw it might seem to be. Theories in the natural sciences frequently describe situations of opposing forces in which the resultant can be calculated only upon very precise quantitative knowledge of all the vectors. A space probe, moving above the earth, may be in a stable orbit, or it may not. Gravity pulls it toward the center of the earth, and its velocity may aim it in quite another direction. Qualitatively, one could say it might crash, or it might escape. A single equation can combine the opposing factors of gravity and momentum, and if accurate numbers can be put in, it provides a definite prediction. Similarly, with religious mobilization, the relative balance of competing forces determines the results.

Under conditions of religious freedom, in which the state does not suppress religious groups of the kinds that are popular in a free market of denominations, a community is likely to have the number of denominations desired by its population—to have reached an equilibrium in its religious market. Factors may work against this principle, notably the excessive cost of supporting a large number of different churches in a community with a small absolute population or the time lag in adjusting to major social changes. But if each community has the number of denominations it deserves, perhaps ultimately reflecting its socioeconomic and ethnic diversity, then diversity may not correlate with mobilization across communities, even if the diversity-mobilization theory is correct. Only if the system is in disequilibrium will the effect appear in ecological data.

Not all forms of disequilibrium will reveal a particular real effect that is concealed by equilibrium. If the need for religious diversity drops, for instance because of ethnic assimilation, then the number of denominations will remain too high, perhaps for a long time. This factor alone would distort the diversity-mobilization correlations. And if both the diversity-mobilization and monopoly-mobilization theories are true, then the result will be a reduction of the proportion of the population involved in churches and apparent evidence only for the monopoly theory.

One would expect ethnic diversity to be a prime determinant of the number of religious denominations in a community, although the evidence in favor of this apparently self-evident idea is weak at best (Greeley 1972; Blau, Land, and Redding 1992). It is possible that ethnic divisions in a population are so nearly coterminous with other divisions that a distinctive ethnic factor cannot be discerned. For example, if ethnic groups are concentrated in different geographic parts of the community, their different needs may be met by several neighborhood churches of a single denomination, each with a distinctive style. This has certainly often been the case with Catholicism in the United States, where separate Irish and Italian congregations exist in different neighborhoods of the same city. If a society is stratified by ethnicity, then separating the population into different denominations by social class simultaneously separates them by ethnicity. Ethnic assimilation, under these conditions, might have no effect on the number of denominations in a community.

The diversity-mobilization theory states that religious monopoly fails to meet the religious needs of some groups in society, and that some degree of religious pluralism will achieve a higher rate of religious involvement. But what is the mathematical function here? Is it possible that two denominations is a sufficient number to meet the varying needs of socioeconomic groups in an ethnically homogeneous society? Or will three groups achieve a much higher degree of mobilization than two? Clearly, there is the logical possibility that a positive mathematical function exists between the number of denominations and mobilization, but it achieves its maximum very quickly. If so, the real amount of variation among communities may lie largely beyond the range where the diversity-mobilization effect manifests itself. Another way of stating this troublesome possibility is to say that current levels of diversity may be so much above true equilibrium levels in many communities that variations in diversity have no implications for mobilization.

Disequilibrium in the United States and Canada is greatly the result of geographical migration and migration rates have strong, robust negative correlations with rates of church membership (Bainbridge 1990). Areas into which large numbers of people are moving tend to have relatively low church-member rates. To some extent, these places may be poor in denominations, as well, but it is also possible that they have substantial numbers of denominations, many of which are poor in clergy and church buildings relative to the population of the community and are thus unprepared to absorb the rush of in-migrants. Few metropolitan areas experience actual population declines, so out-migration may actually be compatible with equilibrium. However, geographical movement of different ethnic groups and social classes within a city often strands older churches where an appropriate clientele is lacking, thus giving a false impression of denominational diversity.

These observations raise the issue of denomination policy. To some extent, denominational diversity is the result of decisions made by the leaders of major denominations, and it is difficult to predict what social factors related to key variables in the mobilization theories might influence those decisions. Despite all the publications in the past fifteen years employing estimates of church membership, one very seldom sees analyses based on a readily-accessible variable from the U.S. Census: the proportion of employed persons who are clergy. In my own experience, this variable does not seem to correlate with anything and I have never discovered anything worth publishing about it. One hesitates to state any conclusions in the absence of serious research studies, but one implication is that many areas of low church membership are not low in clergy and actually may be oversupplied in churches, whatever the number of denominations may be.

Some large liberal Protestant denominations are reluctant to ordain clergy and to send them into an area unless there are congregations waiting for them. This might leave opportunities for smaller and more evangelical denominations, who will rush into areas that are poorly served by mainline denominations. The result may be a high level of religious diversity, even as the failure of some major denominations to recruit actively leaves the church-member rate low.

The existence of one or two highly successful religious movements in a community, ones that accomplish very high levels of mobilization for idiosyncratic reasons having nothing to do with the hypotheses under consideration here, may produce a high church-member rate while driving weaker denominations out of the local market. Thus, diversity may increase mobilization of the general population, while mobilization by one or two particular denominations reduces diversity. The net effect could be a statistical correlation of zero, despite the fact that it results from two highly significant effects.

People may not be especially free to adopt the religious affiliations that would suit them best as individuals, because they are socially tied to other individuals with somewhat different needs. Therefore, to the extent that the social network of a community is highly interconnected or cohesive, the diversity-mobilization effect would be muted. One might also think that cohesion and network interconnectivity strengthen social influence. The relative balance between the diversity and monopoly effects will vary as a function of the relative social cohesion versus social disorganization of the community, and thus variations in the stability and extensiveness of social relationships across time and space may significantly distort empirical research on the sources of religious mobilization.

Perhaps the greatest current debate in the sociology of religion, the secularization controversy, can be conceptualized in terms of diversity and mobilization. Is religion getting stronger or weaker? If it is changing in strength, in response to other social and cultural changes, then one would think that the rate of change is uneven across communities. It is hard to predict how differential secularization might distort empirical tests of the mobilization theories. To some extent, however, standard secularization theories dovetail with monopoly-mobilization theory. One reason people may lose faith in their religion is through exposure to many very different creeds that appear to contradict each other, and "widespread loss of faith" is how many social scientists might define secularization.

The Stark-Bainbridge theory, in contrast, argues that secularization is not confined to contemporary society, but is a universal process found in all societies and eras. The central, low-tension religious organizations of a society are forever being co-opted by secular powers and are constantly suffering erosion of faith. But this weakening of low-tension denominations is offset by the emergence of sect movements that revive their shared religious tradition. With the passage of time, many successful sects moderate and become prey to secularization themselves, but this merely sets the stage for yet more sects to emerge in further schisms. Over very long periods of time, entire traditions can become secularized and the result is stimulation of religious innovation. Out of the myriad of cults that arise in the ruins of an old religion, a very few will thrive, and one or two will establish entirely new traditions, thus starting the cycle all over again.

Where the United States stands in this long-term historical evolution cannot be determined with confidence (Bainbridge 1989, 1993). In their separate work, Stark has tended to stress revival of existing traditions, while Bainbridge has stressed innovation. But in their collaborative work they leave entirely open the question of whether the end of the twentieth century is a period of equilibrium sustained by revival or the beginning of a period of disequilibrium in which old traditions are weakening yet innovation has not yet established a successor tradition. Whether equilibrium or disequilibrium would be better for the researcher is a moot point, not only because we cannot confidently decide which kind of period our society has entered, but also because each brings its distinctive theoretical and methodological problems.

METHODOLOGICAL CHALLENGES

Sociological research is a difficult undertaking, our successes are never complete, and we often must make do with relatively weak standards of inference in using data to test theories (Collins 1989; Blalock 1989; Lieberson 1992). Although we dare not say so too loudly, this means that good empirical studies are almost never conclusive, unless multiple studies performed with diverse methodologies by different teams agree with each other to a high level of consistency. A single good empirical study is at best a plausibility argument, illustrating a theory and suggesting that it is worth taking seriously. Unfortunately, one senses that a necessary (but by no means sufficient) quality for publication in a scientific journal is the expression of great confidence in one's results. Ambiguous studies whose researchers are very dubious about their own results are almost never published.

Some years ago, Laurie Russell Hatch and I did a study testing the hypothesis that the strength of religion in a community, operationalized as the church-member or adherent rates, would correlate negatively with the proportion female in the elite professions of banker, lawyer, physician, and industrial manager (Bainbridge and Hatch 1982). Our three datasets employed church-membership rates based on different data sources: the 1926 census of U.S.

religious bodies (Bureau of the Census 1930), a nongovernmental survey of American denominations carried out in 1971 (Johnson, Picard, and Quinn 1974), and the 1971 census of Canada (Statistics Canada 1974). To our consternation, the results were ambiguous. Somewhat lamely, we suggested that the much stronger support for the hypothesis in the Canadian data reflected the more conservative role of religion in that country. Coefficients wavered across professions and years in a most perplexing fashion, both with and without plausible statistical controls. At a scholarly conference, just after the article was published, a colleague expressed both surprise and pleasure that a journal editor would publish such unclear results.

To some extent, of course, each occupation in the dataset had its own unique characteristics, possessing its own distinctive career paths, barriers to female employment, and routes around those barriers. All these may vary geographically across the communities and some of those variations may correlate with other key variables or with their errors. In addition, all these effects may be powerfully shaped by the histories, recruitment pools, and social networks of particular denominations (cf. Pescosolido and Georgianna 1989). The analogy for mobilization research is that denominations may need to be analyzed separately, because each has a unique social base, history, recruitment strategy, and sensitivity to diversity or monopoly (Blau, Redding, and Land 1993).

Questions are frequently raised about the validity of the surveys of denominational membership done in 1971, 1980, and 1990, on which some of the mobilization studies rest. Hadaway, Marler, and Chaves (1993) have raised the possibility that actual church attendance has dropped in recent decades, even as self reports of attendance have remained roughly constant. Their own evidence concerns a marked discrepancy between claims of church attendance by a random sample of Protestants in one Ohio county with counts of actual attendance at the same churches, all data coming from 1992, so the researchers really do not have evidence of a decline in church membership. But their suggestion is worth taking seriously, and if the gap between the survey data and reality has changed markedly over time, one would think the errors vary also across geographic distance and across various community-level socioeconomic and cultural variables as well.

All those who have worked with the early twentieth-century censuses of religious bodies have been impressed by the apparent quality of the data. Unlike the recent studies that asked central denominational offices to provide county-level membership estimates, the old religious censuses sent a questionnaire to each local church in the nation. Few modern data collection efforts in the sociology of religion have been done with such great care and none have been on so large a scale. However, errors undoubtedly exist and it is very difficult to estimate them.

The church membership for a given city or county frequently includes some unknown number of persons who live outside the geographical unit, perhaps just across the county line, and thus will be counted in the numerator but not the denominator when calculating the church-member rate. The census reports were quite explicit that denominations differed greatly in whether they counted children as members or restricted membership to confirmed adults, and corrections for this problem are undoubtedly less than perfect (Bainbridge and Stark 1981). The rate of geographical migration, in and out of the community, presumably correlates with the error in the membership estimates. All of these errors may vary systematically across the units of analysis. For example, the tendency of a single church or denomination to miss-estimate its membership will cause less relative error in communities with many churches and denominations.

Courageously, Finke and Stark have taken the data series far back to the time of the American Revolution and it is greatly to their credit that they have done so. However, in some of the research quantitative sociology has climbed far out on a limb that may not be strong enough to sustain some of the statistical analyses that are now being done there. The full trove of data employed by Finke and Stark is sufficiently varied, that I tend to accept their model of steadily increasing church membership over the full sweep of American history, and a number of their analyses of particular denominations seem rather convincing to me, as well. But the more estimating and interpolating there is, and the greater the use of control variables, the more uncertainty I feel in the results.

Beginning in 1850, the U.S. censuses began collecting information directly from each local church on the number of seats for parishioners in their houses of worship and some sociologists have mistaken these for membership counts (Desroche 1971; Whitworth 1975). In 1890, data were collected on both parishioners and seats—counting both heads and tails, as it were—and this permitted Finke and Stark (1986) to develop a very clever statistical technique for estimating membership on the basis of seats. Blau, Land, and Redding (1992) adapted this method, in a sophisticated statistical analysis of the expansion of religious affiliation from 1850 to 1930 that made extensive use of interpolation. My own rather thorough familiarity with the data leaves me quite undecided about the appropriateness of concatenating so many assumptions and estimation techniques, and perhaps merely as a matter of personal timidity I have not dared to join the empirical battles described here.

Many sociologists decry all such research on the grounds that quantitative studies of geographically-based rates suffer from the ecological fallacy (Robinson 1950; Hannan 1971). That is, whereas the empirical data have all been at the most macro level, the theory really concerns group processes that mediate between the individual and the community. To a great extent, this so-called fallacy is just a variant of spuriousness and can be handled by a combination of appropriate control variables and caution when interpreting robust correlations (Bainbridge 1992). Correlations in the religious mobilization literature, however, appear far from robust and application of different control variables and statistical procedures appears to give wildly different results.

Few sociological theories are stated in quantitative terms and those that are expressed through well-defined functions tend to be very simple. Thus it is not surprising that the diversity and monopoly theories are not stated with sufficient precision that they can readily be combined to make definitive predictions. Computer simulation is a tool that may help improve the rigor with which we are able to derive and connect theoretical propositions (Bainbridge 1987, 1995; cf. Bainbridge and Stark 1984), but this works best for very narrowly framed theories and for the foreseeable future we cannot expect great rigor of theories that have considerable scope. The alternative is to extract the parameters of a general theory from careful measurement of variables in empirical studies, but the host of problems with existing church membership data leaves one with little confidence that this can be accomplished without substantial new data collection efforts.

CONCLUSION

The pluralism-mobilization theory is not new within American sociology. In 1879 and 1880, my great-grandfather, William Folwell Bainbridge, completed a tour of Protestant missions throughout Asia and the Middle East, to develop what he called a "science of missions," essentially a sociology of the factors that determined relative success of attempts at religious conversion. He noted a movement among some missionaries to divide up the world into spheres of influence by single denominations. "But I have observed that, as a rule, those mission stations of whatever church or denomination, which are left entirely by themselves, both for the present and the prospective future, do not show that activity and develop that strength, which are manifested in those mission fields where the presence or imminence of emulation has been felt. It was evident in Yokohama that Presbyterians and Methodists were prompting each other to a larger measure of evangelizing enterprise than either would have commanded with all the responsibility in the hands of a single mission, even though reinforced to the full extent of the other denomination's resources of men and means" (Bainbridge 1882, p. 270). He then devoted several pages to supportive evidence and a consideration of various social processes that might give rise to this diversity-mobilization effect.

Unfortunately, the sociology of religion, like sociology more generally, has tended to ignore nineteenth-century empirical American social science and to sink its roots instead deep into European social thought. One result has been the tendency to invest very little systematic research effort into close examination of the social dynamics that take place within and around religious groups, at the level of single congregations. The existence of aggregate-level data on church membership, available essentially free to any social scientist willing to expend energy coding and computerizing them, has quite appropriately attracted many sociologists of religion. In some cases, results have been robust and easy to interpret. But when they are far from robust and alternate interpretations abound, as in the mobilization literature, one can only conclude that very different kinds of data will be needed.

In terms of group processes, the diversity-monopoly argument has classic dimensions. On the one hand, a free market offers individuals a range of choices, thus maximizing potential satisfaction of demand and inducing the greatest number of people to be involved in the market. On the other hand, to the extent that religious faith is sustained by unanimity of opinion among a person's exchange partners, religious pluralism erodes faith, thus reducing the perceived value of religious goods and minimizing involvement in the market. These rational-choice and social-influence elements are merged in the original Stark-Bainbridge theory, which derived religion from a set of propositions about human needs, the group processes that produce particular faiths, and the formal organizations that stabilize and sustain them.

Let me state an outrageous but appealing scientific law: *Every theory that informed and intelligent social scientists believe to be true, is true.* Thus there are only two questions to be answered through empirical research. First, what is the domain of the theory; that is, when and to what phenomena does it apply? As Walker and Cohen (1985) note, a theory ideally should contain clear statements of the scope or boundary conditions of its propositions, but in fact theories seldom do. Second, what precise mathematical function describes the strength of the effect predicted by the theory? When two theories that competent sociologists believe to be true contradict each other within some portion of their domains, real effects will cancel each other out to some extent, giving an ambiguous or false picture of the truth value of one or both theories.

Thus, we can suggest two valid responses to the present debacle in macro-level research on religious mobilization. First, the problem cries out for collection of new data in studies designed to test the rival hypotheses, and the best approaches would examine the effects simultaneously at the level of individuals, small groups, and the religious economies of communities. Such studies will be costly, but the scientific gains from investment in them will extend far beyond the narrow limits of religious research, because the competing theories are so solidly based upon general models of group process. Second, researchers should recognize that contextual effects can powerfully shape the macro consequences of micro phenomena. If macro-level variables in fact shift the balance between rival diversity and monopoly effects, then the meaning of small group processes depends powerfully upon factors at a higher level of aggregation. Taken together, these two points argue for the importance of real-world research on small groups and for the realization that group processes can be powerfully affected by the larger social conditions surrounding them.

ACKNOWLEDGMENT

The views expressed in this article do not necessarily represent the views of the National Science Foundation or the United States.

REFERENCES

Bainbridge, W. F. 1882. *Around the World Tour of Christian Missions: A Universal Survey*. New York: C. R. Blackall.

Bainbridge, W. S. 1987. *Sociology Laboratory*. Belmont, CA: Wadsworth.

————. 1989. "Religious Ecology of Deviance." *American Sociological Review* 54: 288-295.

————. 1990. "Explaining the Church Member Rate." *Social Forces* 68: 1287-1296.

————. 1992. *Social Research Methods and Statistics*. Belmont, CA: Wadsworth.

————. 1993. "New Religions, Science, and Secularization." Pp. 277-292 in *The Handbook on Cults and Sects in America*, edited by D. G. Bromley and J. K. Hadden. Greenwich, CT: JAI.

————. 1995. "Neural Network Models of Religious Belief." *Sociological Perspectives* in press.

Bainbridge, W. S. and L. R. Hatch. 1982. "Women's Access to Elite Careers: In Search of a Religion Effect." *Journal for the Scientific Study of Religion* 21: 242-255.

Bainbridge, W. S. and R. Stark. 1981. "Suicide, Homicide and Religion: Durkheim Reassessed." *Annual Review of the Social Sciences of Religion* 5: 33-56.

————. 1984. "Formal Explanation of Religion: A Progress Report." *Sociological Analysis* 45: 145-158.

Blalock, H. M. 1989. "The Real and Unrealized Contributions of Quantitative Sociology." *American Sociological Review* 54: 447-460.

Blau, J. R., K. C. Land, and K. Redding. 1992. "The Expansion of Religious Affiliation: An Explanation of the Growth of Church Participation in the United States, 1850-1930." *Social Science Research* 21: 329-352.

Blau, J. R., K. Redding, and K. C. Land. 1993. "Ethnocultural Cleavages and the Growth of Church Membership in the United States, 1860-1930." *Sociological Forum* 8: 609-637.

Breault, K. D. 1989a. "New Evidence on Religious Pluralism, Urbanism, and Religious Participation." *American Sociological Review* 54: 1048-1053.

————. 1989b. "A Reexamination of the Relationship between Religious Diversity and Religious Adherents." *American Sociological Review* 54: 1056-1059.

Bureau of the Census. 1930. *Religious Bodies: 1926*. Washington, DC: U.S. Government Printing Office.

Christiano, K. J. 1987. *Religious Diversity and Social Change*. Cambridge: Cambridge University Press.

Collins, R. 1989. "Sociology: Proscience or Antiscience?" *American Sociological Review* 54: 124-139.

————. 1993. "A Theory of Religion." *Journal for the Scientific Study of Religion* 32: 402-406.

Desroche, H. 1971. *The American Shakers*. Amherst: University of Massachusetts Press.

Finke, R. 1989. "Demographics of Religious Participation: An Ecological Approach, 1850-1980." *Journal for the Scientific Study of Religion* 28: 45-58.

Finke, R. and R. Stark. 1986. "Turning Pews Into People: Estimating 19th Century Church Membership." *Journal for the Scientific Study of Religion* 25: 180-192.

_____. 1988. "Religious Economies and Sacred Canopies: Religious Mobilization in American Cities, 1906." *American Sociological Review* 53: 41-49.

_____. 1989a. "How the Upstart Sects Won America: 1776-1850." *Journal for the Scientific Study of Religion* 28: 27-44.

_____. 1989b. "Evaluating the Evidence: Religious Economies and Sacred Canopies." *American Sociological Review* 54: 1054-1056.

_____. 1992. *The Churching of America: 1776-1990.* New Brunswick, NY: Rutgers University Press.

Greeley, A. M. 1972. *The Denominational Society.* Glenview, IL: Scott, Foresman.

Hadaway, C. K., P. L. Marler, and M. Chaves. 1993. "What the Polls Don't Show: A Closer Look at U.S. Church Attendance." *American Sociological Review* 58: 741-752.

Hannan, M. T. 1971. *Aggregation and Disaggregation in Sociology.* Lexington, MA: Lexington.

Homans, G. C. 1974. *Social Behavior: Its Elementary Forms.* New York: Harcourt Brace Jovanovich.

_____. 1984. *Coming to My Senses.* New Brunswick, NJ: Transaction.

Johnson, D. W., P. R. Picard, and B. Quinn. 1974. *Churches and Church Membership in the United States.* Washington, DC: Glenmary Research Center.

Land, K. C., G. Deane, and J. R. Blau. 1991. "Religious Pluralism and Church Membership: A Spatial Diffusion Model." *American Sociological Review* 56: 237-249.

Lieberson, S. 1992. "Einstein, Renoir, and Greeley: Some Thoughts about Evidence in Sociology." *American Sociological Review* 57: 1-15.

Pescosolido, B.A. and S. Georgianna. 1989. "Durkheim, Suicide, and Religion: Toward a Network Theory of Suicide." *American Sociological Review* 54: 43-48.

Robinson, W. S. 1950. "Ecological Correlation and the Behavior of Individuals." *American Sociological Review* 15: 351-357.

Simpson, J. H. 1990. "The Stark-Bainbridge Theory of Religion." *Journal for Scientific Study of Religion* 29: 367-371.

Stark, R. and W. S. Bainbridge. 1985. *The Future of Religion.* Berkeley: University of California Press.

_____. 1987. *A Theory of Religion.* New York: Toronto/Lang.

Statistics Canada. 1974. *1971 Census of Canada.* Ottawa: Queen's Printer.

Walker, H. A. and B. P. Cohen. 1985. "Scope Statements: Imperatives for Evaluating Theory." *American Sociological Review* 50: 288-301.

Warner, R. S. 1993. "Work in Progress toward a New Paradigm for the Sociological Study of Religion in the United States." *American Journal of Sociology* 98: 1044-1093.

Whitworth, J. M. 1975. *God's Blueprints.* London: Routledge and Kegan Paul.

GENDER DIFFERENCES IN REFERENCE GROUPS, SELF-EVALUATIONS, AND EMOTIONAL EXPERIENCE AMONG EMPLOYED MARRIED PARENTS

Robin W. Simon

ABSTRACT

This paper first reviews the conceptual and empirical work on reference groups and suggests ways in which attention to reference groups, and the self-evaluations based on such groups, could enhance theory and research on gender differences in the psychological consequences of multiple role occupancy. To illustrate the potential importance of social comparisons in the stress process, the paper then examines gender differences in reference groups, self-evaluations, and emotional experience among employed married parents. Based on data from follow-up, in-depth interviews with 40 individuals who participated in a community panel study of mental health, I find that men's and women's evaluations of themselves as spouses and parents vary. I suggest that differences between men's and women's self-evaluations are traceable to the different groups they select as their frame

Advances in Group Processes, Volume 12, pages 19-50.
ISBN: 1-55938-872-2

of reference. Women tended to compare themselves to their mothers or other nonemployed women who were more involved than they were in family life. Social comparisons were associated with negative self-evaluations (as wives and mothers) and negative feelings of inadequacy, self-doubt, and guilt. In contrast, men tended to compare themselves to their own fathers and other employed males who were less involved than they were in family life. Social comparisons were associated with positive self-evaluations (as husbands and fathers) and positive feelings of pride, self-satisfaction, and self-worth. By identifying gender differences in reference groups, self-evaluations, and emotional experience among employed married parents, this paper provides further insight into why the psychological benefits of combining work and family roles are greater for men than for women. The utility of the concept of reference groups, and the self-evaluations based on such groups, for future theory and research on gender, social roles, and mental health is discussed.

INTRODUCTION

Reference groups, and the social comparisons that are based on such groups, are important sources of information about the self. Because social comparisons provide self-relevant information, the *selection* of reference groups has important consequences for self-evaluations and feelings of inadequacy or self-worth. In this paper, I first review the conceptual and empirical literature on reference groups and discuss ways in which attention to reference groups, and the self-evaluations based on such groups, could enhance theory and research on gender differences in the psychological consequences of multiple role occupancy. I then provide a few examples from in-depth interviews that highlight the relationships among reference groups, self-evaluations, and emotional experience among employed married parents. In the final part of the paper, I discuss the utility of the concept of reference groups, and the self-evaluations based on such groups, for future theory and research on gender, social roles, and mental health.

CONCEPTUAL AND EMPIRICAL WORK ON REFERENCE GROUPS AND SOCIAL COMPARISONS

The concept of the reference group has had a relatively long history in sociology. A reference group refers to either a social organization or an individual which social actors employ as a basis for self-knowledge and self-evaluation. The basic idea underlying the concept of the reference group is that individuals routinely engage in social comparisons with a group or an individual in order to interpret whether their social situations, values, and behavior (e.g., their role performances) represent successes or failures (Hyman 1942, 1960).

Early conceptual work on reference groups indicates that there are two distinct, though related, ways in which individuals use reference groups (Kelley 1968). The first way individuals use reference groups is as a *standard of comparison* against which they could evaluate themselves. In this situation, reference groups serve as a frame of reference, or a yard stick, for making judgments about oneself and assessing the adequacy of one's beliefs and role performances. In addition to their comparative function, individuals also use reference groups to *ascertain norms* for role-related behavior. In this case, reference groups represent standards for how an actor should or ought to think, feel, and behave in social situations.[1] Reference groups are thought to be particularly important sources of comparative and normative information when clear norms and objective standards for self-appraisals are unavailable (Festinger 1954). Moreover, while people may invoke different reference groups for comparative and normative purposes, a single individual or group is commonly used for both of these functions. For instance, the same group may serve as a standard against which individuals evaluate the adequacy of their role performances as well as the source of their norms and attitudes. Some have argued that social comparisons have consequences for self-evaluation only to the extent that the standard of performance constitutes an expectation that the individual feels *should* be met (Hyman and Singer 1968; Singer 1981).

Theory and research on reference groups have focused on two analytically distinct, though related, aspects of reference group behavior; these include the *selection* of comparison groups and the *consequences* of social comparisons for attitudes and conduct. Overall, sociologists have devoted more attention to the effects of social comparisons than to the determinants of reference groups.

With respect to the selection of reference groups, it has been suggested that people choose, as their referents, others who are similar or close to themselves in one way or another (e.g., in sex, age, race, class, social status, social roles, group membership, ability, and proximity [Festinger 1954]). Because of their emotional importance, family and other primary group members are often selected as frames of reference. Since people typically hold multiple statuses and roles, it is likely that they have numerous reference groups of varying importance. The little empirical sociological research that exists on this aspect of reference group behavior has not provided a clear understanding of what determines the selection of reference groups, although the structure of situations, social norms, as well as aspirations for group membership all seem to be important factors (Form and Geschwender 1962; Hyman 1942;

Rosenberg and Simmons 1972; Rosow 1967; Stern and Keller 1953; Strauss 1968).

With regard to the consequences of reference groups, sociologists have focused on either the effects of normative reference groups for conformity, or the consequences of comparative reference groups for self-evaluation.[2] A central idea is that normative reference groups play a role in socially shaping individuals' attitudes and behaviors, and are instrumental at producing attitudinal and behavioral change. To the extent that attitudes and behavior are organized around a particular social status or role, a consequence of social comparisons is socialization (Singer 1981). Scholars interested in the effects of normative reference groups have sought to specify the conditions that impede or facilitate conformity. For instance, in his classic study of attitudinal change, Newcomb (1943) found that college student's desire for membership in a sorority or fraternity facilitated their adoption of (i.e., their conformity to) the reference group's values and behaviors. This and other similar findings (Eisenstadt 1968) suggest that the psychological salience, or importance, of an existing or a desired identity for one's self-conception would increase the influence of the reference group. To date, however, the degree to which the relative salience of identities *moderates* the influence of normative reference groups has not been explored in much detail.

While the work on normative reference groups focuses on their effects for conformity, another line of work emphasizes the effects of social comparisons for self-appraisals. Several authors have noted that reference groups, and the social comparisons that are based on such groups, are important sources of self-evaluation and self-esteem and are, therefore, central to the self-concept (Rosenberg and Simmons 1972; Rosenberg and Pearlin 1978). Because individuals learn about themselves primarily through reflected appraisals and social comparisons, the self in part depends on reference groups (Suls and Miller 1977). Insofar as social comparisons are a source of self-evaluation and feelings about oneself, the choice of reference groups would be particularly important for those aspects of self that are perceived by the individual to be highly salient. However, the implications of either positive or negative social comparisons for highly salient versus non-salient identities have not been elaborated in the existing sociological work.[3]

Interestingly, although the selection and consequences of reference groups have been conceptualized as analytically distinct phenomena and processes, the most interesting finding from the corpus of research on reference group behavior is that the *effects* of social comparisons depend, in large measure, on the *choice* of the reference group. Numerous studies have documented that feelings of satisfaction or dissatisfaction with one's self or situation, and feelings of self-worth or inadequacy, depend neither on the situation itself nor on the objective characteristics or circumstances of individuals. Rather, social and self-evaluations depend on the individual's comparison between themselves and some other group or individual (Crosby 1976; Hyman 1942; Merton and Rossi 1950; Rosenberg and Simmons 1972; Strauss 1968). In other words, self-appraisals and subsequent feelings of self-esteem appear to be *contingent* on the particular reference group selected as a frame of reference.

For instance, in the first explicit discussion of reference groups, Hyman (1942)[4] showed that people's perceptions of their social position depended on the particular group they used as a framework for comparison, rather than on the objective indicators of their status (e.g., their education and income). People who were experimentally assigned high status comparison groups perceived themselves as having lower status than they actually had, whereas those assigned to low status groups perceived themselves as having relatively high status. Hyman further showed that changes in judgment about one's social position could be brought about by experimentally changing the status of the reference group. Similar findings are also evident in nonexperimental situations across a range of social phenomena.

Merton and Rossi (1950) showed that World War II soldiers' feelings of satisfaction with their situations depended on the groups they compared themselves to, and not on the objective characteristics of their situations (such as how close they were to combat areas). Similarly, Crosby (1976) found that despite their relatively low wages, employed women did not perceive their wages as either low or unfair because they compared themselves not to men, but to other female workers who were equally low paid. Along these same lines, Strauss (1968) reported that blind people were more likely to experience feelings of inferiority and incompetence when their frame of reference was the sighted. However, when blind respondents compared themselves to other blind individuals, they were more likely to perceive themselves as competent social actors. Other studies on different aspects of social life have yielded similar results (e.g., see Easterlin [1973] and Elder [1974] on feelings of economic deprivation; Rosow [1967] on feelings of competence and self-worth among the elderly; Patchen [1961] on job satisfaction; Pettigrew [1968] on perceptions of racial inequality; Parker and Kleiner [1968] on mental illness; and more recently, Felson and Reed [1986] on self-appraisals of academic achievement among children).

One of the most notable examples of this phenomenon can be found in Rosenberg and Simmons' (1972) study of self-esteem among Black and white children. As part of a larger project on the effects of school desegregation, Rosenberg and Simmons found that Black secondary school students in integrated settings had lower self-esteem than those in segregated settings because they compared themselves to white children in their school who were comparatively better off. Ironically, Black children in segregated middle schools had relatively high self-esteem because they compared themselves to other Black children who were as, or more, disadvantaged. In short, these findings indicate that the consequences of social comparisons for global self-esteem depend in large measure on the particular group which is employed as the frame of reference.

On the basis of both experimental and nonexperimental research, it thus appears that feelings of satisfaction or dissatisfaction with one's self and social situation, and feelings of self-worth or inadequacy, are contingent on the selection of the reference group. When individuals compare themselves to others who are worse off than (or not doing as well as) themselves, the result is satisfaction and self-enhancement (i.e., a positive self-evaluation). Conversely, when social comparisons are based on others who are better off (or doing better than) themselves, the result is dissatisfaction and self-depreciation (i.e., a negative self-evaluation) (Singer 1981). While the implications of positive evaluations have not been fully specified, Crosby (1976) noted that negative evaluations may lead either to striving for achievement or emotional and/or physical symptoms of stress, depending on the individual's sense of control to change his or her self or social situation.

Although not itself a theory, the importance of reference groups and social comparisons have been acknowledged in numerous social psychological theories. For example, in addition to social comparison theory (Festinger 1954; Merton and Rossi 1950; Suls and Miller 1977), theories about socialization (e.g., symbolic interactionism and social learning theory), deviance (e.g., differential association and control theory) and social influence, role theory (Turner 1956), equity theory (Walster, Walster, and Berscheid 1978) and its offshoot, relative deprivation theory (Crosby 1976) all emphasize the role reference groups play in socially shaping individuals' attitudes and behaviors. Theories about the self-concept have also recognized the impact social comparisons have on the development and maintenance of self-concept and self-image (Rosenberg 1979; Rosenberg and Simmons 1972; Rosenberg and Pearlin 1978). Interestingly, while these theories highlight the contributions of reference groups to a broad range of social psychological phenomena, sociological theories about the mental health effects of social roles have generally not considered the potential importance of social comparisons in the stress process. Nor have these theories considered the potential links between reference groups, self-evaluations, and emotional functioning. The lack of attention to the psychological consequences of reference groups and social comparisons in theories about social roles and mental health is surprising since scholars have long recognized that reference groups, and the social comparisons based on such groups, are a means through which individuals assess their own self-worth and the adequacy (or conversely, the inadequacy) of their various role performances.

In light of the current ambiguity of norms governing certain major adult social roles, as well as the unclarity of standards for role-related behavior, it is likely that reference groups and the self-evaluations based on such groups are important for understanding gender differences in the emotional effects of multiple role occupancy. In fact, an examination of the reference groups men and women employ as a basis for social comparison and self- evaluation may provide insight into why combining work and family roles is more stressful and less protective of the well-being of women relative to men. To date, the only sociologists who have considered the mental health effects of social comparisons and self-evaluations are Parker and Kleiner (1968). In their attempt to explain the higher rates of psychological disorder of Blacks who had recently migrated to northern cities compared to native urban Blacks, Parker and Kleiner hypothesized (and found) that individuals' evaluated themselves as failures if, in an area relevant to their self-esteem, they perceived their performances as falling below that of the reference group. Conversely, people evaluated themselves as successful if, in an area relevant to their self-esteem, they perceived their performances as living up to or surpassing that of their reference groups. Although they focused on the effects of reference groups for self-evaluations of social position (rather than the effects of self-evaluations in *specific* role domains), Parker and Kleiner's work showed that negative discrepancies between one's role behavior and that of their reference group is psychologically stressful, assaults self-esteem, and results in a propensity for psychological disorder.

The well-documented phenomenon of gender differences in mental health among employed married parents provides a unique opportunity to examine the potential links among reference groups, self-evaluations in specific role domains, and emotional functioning. Over the past three decades, epidemiological research has consistently showed that employed married mothers report significantly higher levels of psychological symptoms than employed married fathers.[5] On the basis of the preceding discussion, we could expect that employed husbands' and wives' evaluations of themselves as spouses and parents, and their feelings about combining work and family roles, vary, depending on whether they perceive themselves (and their role performances) as falling short of, or living up to and perhaps exceeding, the standards of their respective reference groups. To the extent that social roles, and the identities based on these roles, vary in their psychological salience, it is reasonable to also expect that negative self-evaluations in family role domains will be particularly troublesome for well-being if the roles and corresponding identities being evaluated are perceived by the individual as central to his or her self-concept (Simon 1992a; Thoits 1991).

THEORY AND RESEARCH ON GENDER DIFFERENCES IN THE PSYCHOLOGICAL CONSEQUENCES OF MULTIPLE ROLE OCCUPANCY

Much of the now extensive literature on the relationship between social roles and mental health has focused on gender differences in the psychological consequences of multiple role occupancy. Theory and research on this topic indicate that the impact of social roles on men's and women's well-being differs, depending on the specific role combination in question. In light of recent increases in female employment, and subsequent changes in married women's role configurations, it is not surprising that the "breadwinner" role combination (i.e., spouse-parent-worker) has received considerable scholarly attention. Social change in married women's role constellations over the past few decades has been accompanied by a proliferation of both theory and research on the emotional effects of employment for married women.

The initial work on this topic was concerned with assessing whether employed wives are *less* distressed than homemakers, and if employed wives enjoy the *same* mental health advantages as employed husbands. Overall, the findings of studies based on within gender comparisons have been mixed with respect to the psychological consequences of employment (and multiple role occupancy) for married mothers. For instance, although several studies showed that employed wives are less distressed than homemakers (Gore and Mangione 1983; Gove and Geerken 1977; Kandel, Davies, and Raveis 1985; Kessler and McRae 1982; Rosenfield 1980), other studies found no difference between the symptoms of employed wives and homemakers (Aneshensel, Frericks, and Clark 1981; Cleary and Mechanic 1983; Gore and Mangione 1983; Kandel, Davies, and Raveis 1985; Pearlin 1975; Radloff 1975; Roberts and O'Keefe 1981). In contrast to these studies, findings of research that compared the symptoms of employed wives and employed husbands have been more consistent. Several studies have documented that employed wives with children at home are more anxious, somatic, and distressed (and sometimes more depressed) than similar husbands (Cleary and Mechanic 1983; Kessler and McRae 1982; Menaghan 1989; Thoits 1986). Taken together, this research indicates that the mental health advantages of combining work and family roles are greater for men than for women.

While this earlier work was preoccupied with documenting the emotional effects of employment for married mothers, more recent research has been concerned with identifying the factors that are responsible for why the psychological benefits of multiple role occupancy are fewer for women relative to men. A variety of factors residing within as well as outside the family are now recognized as contributing to the gender gap in mental health among employed married parents.

For example, the current division of housework and child care within the family, and husbands' modest contribution to it, results in role conflict and role overload for employed wives which puts them at higher risk of psychological disorder relative to employed husbands (Pearlin 1975; Kessler and McRae 1982; Ross, Mirowsky, and Huber 1983). Husbands' and wives' preferences for the wife's employment versus homemaking are also partially responsible for gender differences in distress among married persons (Ross, Mirowsky, and Huber 1983). Moreover, labor market inequality for women, another feature of contemporary social organization, also contributes to distress differences between employed wives and husbands. A consequence of occupational sex segregation is that women tend to be concentrated in jobs that offer less potential for control and autonomy, job advancement, and personal gratification; all of these job characteristics promote feelings of self-esteem and psychological well-being for both men and women (Haw 1982; Kasl 1989; Lennon and Rosenfield 1992; Link, Lennon, and Dohrenwend 1993; Loscocco and Spitze 1990; Lowe and Northcott 1988; Miller, Schooler, Kohn, and Miller 1979). Wives' relatively low incomes also have implications for mental health as they affect their marital power and sense of personal control vis-à-vis their husbands (Rosenfield 1989, 1992).

Other researchers emphasized the role of cultural factors. The lack of cultural support for married mothers' employment and the relative proportions of employed wives to homemakers have been linked to gender differences in the mental health advantages of employment for married parents. According to Thoits (1986) and Menaghan (1989), the breadwinner role combination is more protective for males than for females because it continues to be a "normative" role situation for men and a "nonnormative" role situation for women. It is interesting that implicit in these authors respective accounts of gender differences in the emotional benefits of multiple role occupancy is the assumption that employed married mothers lack appropriate *role models* of *working* wives and mothers.

Finally, I have argued that gender differences in the meaning of work and family roles, which are rooted in sociocultural beliefs about the interrelationships between work and family for men and women, also contribute to the gender gap in mental health among employed married parents. In a recent qualitative study (Simon 1992b, 1995), I found that men and women believe there is greater overlap, or interdependence, between work and family role obligations for men than for women. This research suggests that married fathers may derive more psychological benefit from employment than married mothers because employment contributes to men's ability to meet their normative family obligations, whereas employment detracts from women's ability to meet their traditional obligations as wives and mothers.

However, while these authors have all identified a variety of role-related factors that help account for distress differences between employed husbands and wives, most scholars working on this topic have neither acknowledged nor explored men's and women's *evaluations* of themselves as spouses, parents, and workers and their *feelings* about combining work and family. The lack of attention to these self and emotion processes is unfortunate, since the breadwinner role combination has different implications for men's and women's self-evaluations and feelings, particularly as spouses and parents. Differences between men's and women's evaluations of themselves as spouses and parents, and their feelings about combining work and family roles, may be additional social psychological factors contributing to the gender gap in mental health among employed married parents.

Because reference groups are an important source of self-knowledge and constitute the basis for self-evaluations and feelings of self-worth or inadequacy, the particular groups or individuals men and women select for social comparisons may provide insight into why combining work and family roles results in different self-evaluations and feelings for men and women. For example, if employed married fathers compare themselves to others who are *less* involved than themselves in family life, social comparisons should result in favorable self-evaluations (as husbands and fathers) and positive feelings (e.g., feelings of self-worth). If, on the other hand, employed married mothers compare themselves to others who are *more* involved than themselves in family life, social comparisons should result in unfavorable self-evaluations (as wives and mothers) and negative feelings (e.g., feelings of inadequacy). It also is possible that if husbands of employed wives compare themselves to husbands of homemakers who are *able* to provide a family wage, social comparisons should result in unfavorable self-evaluations (as husbands, fathers, and providers) and negative feelings for these men. To the extent that family role-identities are important for men's and women's self-conceptions, positive self-evaluations in family role domains and feelings of self-worth should enhance well-being, whereas negative self-evaluations in these role domains and feelings of inadequacy should erode well-being and be troublesome for mental health (Simon 1992a; Thoits 1991).

Given social change in men's and women's roles over the past few decades, and the unclarity of norms underlying role-related behavior, the groups men and women employ as their frame of reference may, therefore, also be involved in why combining work and family roles is more stressful and less rewarding for women relative to men. Since the outcomes of social comparisons for self-evaluation, feelings of self-worth or inadequacy, and emotional well-being may be *contingent* on one's choice of a reference group, it is important to examine the groups or individuals men and women draw on as their frame of reference.

In my research on the emotional effects of multiple role occupancy (Simon 1992b, 1995), I found that combining employment with marriage and parenthood resulted in *different* evaluations for men and women as spouses and parents. Although it was not my intention to examine men's and women's reference groups (and social comparison processes), there was some evidence in my data that suggests that employment and multiple role occupancy resulted in different self-evaluations for men and women as spouses and parents, in part, because they employed *different* individuals and groups for self-appraisals.

In a following section of the paper, I will discuss differences between men's and women's self-evaluations as spouses and parents and their feelings about combining work and family which were evident in my research. I also will provide a few examples from in-depth interviews that suggest that men's and women's evaluations of themselves as spouses and parents, and the emotions they experience from combining work and family obligations, can be traced to the different reference groups they employ for social comparisons. Before I turn to these examples, however, it is important to first briefly describe the larger study, including the data and methods.

DATA AND METHODS

The examples presented in this paper come from a study based on follow-up, in-depth interviews with men and women who had participated in a two-wave prospective panel study of mental health in Indianapolis. Structured personal interviews were administered in 1988 to a representative sample of 354 married individuals (located through random digit dialing) and again in 1990 with 289 located persons. Information about sampling procedures, response rates, attrition, and the characteristics of the panel are reported elsewhere (see Simon 1992a, 1992b; Thoits 1992, 1995). In-depth interviews were conducted in 1991 with a subset of 40 *full-time employed married mothers* and *fathers* who had at least *one child under 18 years in the household* and whose *spouse* also was *employed full-time.* This subset was selected because the purpose of the follow-up study was to explore gender variation in the meaning of social roles for men and women who have the same role configuration and role situation, and to generate hypotheses about the social psychological processes which may help account for sex differences in distress. Since I expected that these social psychological processes vary by race, I restricted the follow-up sample to white respondents, thereby limiting the generalizability of the follow-up study. Eligible respondents were identified by computer generating the case identification numbers of those persons whose characteristics "fit" the sample requirements. Over 90 percent of the eligible contacted persons agreed to be reinterviewed. Although the findings discussed in this paper may be generalizable to other white employed married parents in dual-earner marriages, they are not meant to reflect the experiences of all men and women. Selected sociodemographic characteristics of the follow-up study are reported in the Appendix. This sample included men and women from a range of social class and educational backgrounds. With the exception of age and family income, male and female respondents were similar in sociodemographic characteristics. The men were slightly older than the women and were somewhat more likely to have higher household incomes.

The in-depth interviews included a series of questions designed to tap respondents' beliefs about the obligations underlying their social roles. To examine respondent's self-evaluations in family role domains, they were asked to evaluate how "good" a spouse and parent they thought they were, and if being employed "adds to," or "takes away from," their ability to be the type of spouse and/or parent they would like to be. Additional role performance data were obtained when respondents described their feelings about combining employment with marriage and parenthood as well as their feelings about having a spouse who combines work and family. To further explore gender variation in self-evaluations, I asked respondents to compare their own feelings about combining multiple roles with their perceptions of their spouse's feelings about combining work and family. Probes were used in conjunction with each question in order to obtain detailed and specific answers. Tape recordings of each interview were transcribed and themes were content coded. Computer searches were conducted on the codes in order to identify all references to information relevant to reference groups, social comparisons, self-evaluations, and feelings.

Three points about the analyses are noteworthy: First, because I did not question respondents about their reference groups, it is likely that my findings *under*estimate both the implicit and explicit social comparisons men and women make for self-appraisals. The inclusion of questions on reference groups may also have revealed more *within* gender variation in reference groups, especially among women, than my data suggest. Second, while the paper highlights the importance of men's and women's *choice* of reference groups, my data do not allow me to examine the processes through which selection occurs. Relatedly, although I suggest that differences between men's and women's reference groups *contribute to* their different self-evaluations as spouses and parents, it is equally possible that men and women choose reference groups that allow them to *maintain* and/or *confirm* preexisting self-evaluations and feelings. Although it is important for future research to sort out the causal direction of this relationship, the purpose of this paper is both to illustrate the associations among reference groups, self-evaluations, and emotional experience as well as to develop hypotheses about their potential consequences for stress and mental health.

GENDER DIFFERENCES IN SELF-EVALUATIONS AND EMOTIONAL EXPERIENCE: MEN'S AND WOMEN'S REFERENCE GROUPS AND SOCIAL COMPARISONS

The in-depth interviews revealed that employment and multiple role occupancy have different effects on men's and women's evaluations of themselves as spouses and parents as well as on their feelings of adequacy and self-worth. For almost all of the men (90%), combining work and family obligations resulted in positive self-evaluations as fathers and husbands and positive feelings. In contrast, holding work and family roles resulted in negative self-evaluations as mothers and wives and negative feelings for over two-thirds (70%) of the women. I suggested that employment and multiple role occupancy have different effects on men's and women's self-evaluations in family role domains because wives' employment means they are not continuously available to their spouse and children for emotional support and nurturance, whereas husbands' employment means they are partially fulfilling their roles as husbands and fathers. I interpreted these findings as suggesting that sex differences in role meaning and self-evaluations are important mediating variables in the production of distress differences between employed wives and husbands (Simon 1995).

Yet, differences between the groups men and women employ as their frames of reference also appeared to be implicated in gender differences in self-evaluations and the emotional benefits of multiple role occupancy. In the process of conducting the interviews (and later, analyzing the data), it became evident that when assessing themselves as spouses and parents, several respondents (N = 28) *spontaneously* compared themselves to their own parents and/or same sex peers. In some cases, social comparisons were quite explicit, while in other cases they were more covert and implicit. These data suggest that men's and women's evaluations of themselves as spouses and parents, and their feelings about combining work and family (e.g., their feelings of self-worth or inadequacy), are associated with the different groups they employ as their frame of reference.

The following examples from the in-depth interviews illustrate the links between reference groups, self-evaluations, and emotional experience among employed wives and husbands. From these examples, it is clear that men's and women's reference groups, which often times consisted of childhood role models, served not only as a *standard for social behavior*, but also as a *source of norms* regarding gender specific family role behavior. The groups men and women selected as frames of reference not only represented, but also embodied, sociocultural beliefs about the traditional family role obligations of males and females.

Women's Reference Groups: Employed Married Mothers' Negative Self-Evaluations as Wives and Mothers and Feelings of Inadequacy

Overall, combining employment with marriage and parenthood resulted in negative self-evaluations for the majority of women. As indicated earlier, 70 percent (N = 14) of the women interviewed indicated that they often feel inadequate as wives and mothers because they are employed. A theme which emerged from the female data was that women think their employment interferes with their ability to adequately fulfill their family roles which, for them, involves the provision of *round-the-clock* emotional support and nurturance. According to the women in the sample, employment was a threat to their identity as a "good" mother and wife because it prevented them from "being there for" (i.e., spending *enough* time with) their children and husbands. The emotions these women experienced from combining work and family obligations were consistent with their self-evaluations of performances in family role domains. In addition to feelings of inadequacy, these women also felt guilty because they perceived that a consequence of their jobs and their multiple role involvements is that they do not give their children and husbands the attention they need.

When I probed their responses and asked them why they experienced these negative feelings and emotions, several of these women (N = 11) responded by comparing themselves to other groups or individuals. Interestingly, the groups to which these women compared themselves tended to consist of *non*employed females, most notably their own mothers or other homemakers, who were *more* involved than they were in family life. By comparing themselves to more "traditional" wives and mothers, these women perceived that their own family role behavior *fell short* of the behavior of their reference group.

For example, although this 40-year-old divisional assistant said she no longer feels guilty about combining employment with parenthood because her children are older, she reflected on an earlier period in her life when she experienced this negative emotion.[6] When I asked her why employment resulted in negative emotions when her children were younger, she explained that being home with her children (rather than at work) was what she thought society *expected* of her.

> From a guilt point of view, [it was] what I thought society expected of me. That
> a mother should be home with her children. (#0022F)[7]

When I continued to probe and asked her why combining work and parenthood
made her feel guilty when her children were younger, she immediately referred
to her own childhood experiences of growing up in a family with a *stay-at-home* mother.

> Probably the biggest thing was that I didn't grow up that way. My father worked
> and my mother was always at home, so I didn't grow up in an environment where
> both parents worked. That was probably the biggest thing, the single most
> important way I looked at things, because I think the way a person grows up
> has a lot to do with the way they view life. And so [the reason] why I felt guilty
> [was] because it wasn't the way I grew up. (#0022F)

For several other women, combining work and family roles resulted in
feelings of inadequacy. For example, a 27-year-old bookkeeper who was
pregnant with her second child mentioned that she often feels inadequate as
a mother because she is unable to care for her child during the day. She
described her current feelings and also projected the negative emotions she
expects to feel when she returns to work after her second child is born.

> It makes me feel like I am failing him [and that] I am not being a good mother.
> I think I'm going to feel the same way when this one is born. It's going to help
> that the baby [will be] with my sister-in-law, but she still isn't going to give him
> the care that I want to give him. She babysits for several other children, so he's
> not going to have the individualized time that I could give him [by] having only
> two at home. She's got ten other babies that she babysits for. They're all relatives,
> but... (#0980F)

Interestingly, although this respondent's own mother was herself employed
when she was growing up, she nevertheless compared herself to an earlier
cohort of nonemployed mothers and wives who were able to spend more time
taking care of their children, their husbands, and their homes.

> Well, I think it [has to do with] the way you're brought up. I think with the
> generation today, the way we grew up, there were so many more women who
> didn't work. They were able to keep the house spotless, keep all the laundry done,
> take care of the kids, have dinner on the table. So it's like, we're suppose to be
> able to do all of that plus we're suppose to be able to have a full-time job and
> keep our husbands happy. (#0980F)

In addition to feelings of guilt and inadequacy, several women also talked about their feelings of *self-doubt*. For example, a 33-year-old sales manager described her feelings of uncertainty and mentioned that she copes with her feelings of self-doubt by decompartmentalizing her various role obligations (see Stryker and Statham 1985).

> [I wonder] who's suffering. Am I spending enough time with my kids? Am I doing the right thing? Would my kids be better off with less of a lifestyle, less of a neighborhood, and more time with me? Or are they better off growing up in this environment? It's an overwhelming sense. I pretty much have found [that] in order to handle everything, you have to break everything down. You really can't look at the whole picture. [When] you look at the whole picture, you start to wonder how in the hell are you doing it and then you start having doubts. Am I doing it okay? Who's suffering? You really have to break it down and take one thing at a time, in my opinion anyway. I can't handle the whole thing as one ball of wax. (#0473F)

When I probed her response and asked her why she has feelings of self-doubt as a mother and concerns that her children may be "suffering," she attributed those negative feelings and emotions to the fact that her mother did not work outside the home and was "there" for her when she was growing up.

> I think it's pretty much the way we were brought up. You know, mom was pretty much it. Mom didn't work. Mom was there. (#0473F)

In short, employment and multiple role occupancy resulted in negative self-evaluations for many women as mothers and wives and negative feelings such as self-doubt, inadequacy, and guilt. The examples presented here suggest that the groups and individuals women selected as their frame of reference may have contributed to their negative self-appraisals and emotions. In the absence of clear norms and objective standards for evaluating their various role performances, employed wives appeared to engage in social comparisons with other *women*. When evaluating themselves as wives and mothers, several women spontaneously compared themselves to their own mothers and other nonemployed wives, who they perceived as being more active than they were in family life. By choosing as their reference group an earlier generation of traditional homemakers, social comparisons led to perceptions of their own inadequacy as wives and mothers, since their family role behavior did not live up to the *standard* of their reference group. However, it also appeared that women's reference groups served not only as a standard for family role behavior, but also as a source of norms regarding how they *ought* to act as wives and mothers. In addition to experiencing negative feelings of inadequacy and self-doubt, these women also felt guilty toward their children and husbands because they violated the norm for women of continuous emotional support and availability, which was a norm embodied by their reference group.

In light of the centrality of family roles for women's self-conception, it is not surprising that employment and multiple role occupancy are both stressful and distressing for them. To the extent that family roles continue to be an important source of self-definition for employed women, it is reasonable to expect that unfavorable social comparisons as wives and mothers have negative consequences for their mental health. Negative self-evaluations and a configuration of negative emotions (including feelings of self-doubt, inadequacy, and guilt) in these highly valued role domains—which, in part, reflect the groups employed wives select as their frame of reference—undermine their feelings of self-worth, which in turn, may contribute to their relatively high symptom levels.

Men's Reference Groups: Employed Married Fathers' Positive Self-Evaluations as Husbands and Fathers and Feelings of Self-Worth

In sharp contrast to the women, employment (and multiple role occupancy) resulted in positive self-evaluations for the majority of men. Recall that 90 percent (N = 18) of the men I interviewed indicated that they often feel successful as fathers and husbands because they are employed. A theme which emerged from the male data was that men think their employment contributes to their ability to successfully fulfill their family roles which, for them, involves the provision of financial support among other things. According to the men in the sample, employment bolstered their identities as a "good" father and husband because they were able to *provide*. In a parallel manner to the women, the emotions these men experienced from combining work and family obligations were consistent with their evaluations of their performances in family role domains. In addition to feelings of success, employment and multiple role occupancy resulted in positive emotions such as pride, self-satisfaction, and a sense of self-worth for men.

When I probed their responses and asked them why they experienced positive feelings and emotions from combining work and family, some of these men (N = 10) responded by comparing themselves to other groups or individuals. Interestingly, and again in a parallel manner to the women, the groups to which these men compared themselves tended to consist of other *employed* males, most notably their own fathers and/or male peers, who were considerably less involved than they were in family life. By comparing themselves to more traditional *"noninvolved"* fathers and husbands, these men perceived that their own family role behavior not only lived up to, but actually exceeded, the behavior of their reference group.

For example, although this 50-year-old systems associate worked relatively long hours at his job, he nevertheless evaluated himself as a "good" father. When I probed his response and asked him why he thought he was successful in this role domain, he responded by comparing himself to his own father who was less involved with him (when he was growing up) than he currently is with his own children.

> I think I'm a pretty good father. I think I could be better. You can always do a better job, but I think I'm a good father. I say that because my two children have never caused any problems, have never been in any kind of serious situation from being in trouble to being delinquent. I would think that's a result of the kind of things that they've learned from both my wife and myself. I look back at my childhood and I think my father didn't spend very much time with me, and again that was not because he didn't love me. He did. He had to work and it was a different situation. Being raised back in the forties, my father had to work two jobs. As a result, [he] was not able to do a lot of things [with me such as] attend school things, or [be involved] when I was active in basketball, and so on. So I think I'm a good father in that I do spend time with my children. And we've done things together. (#1086M)

In addition to their positive self-evaluations as fathers (and husbands), several men also talked about their feelings of *pride* from combining work and family obligations. A 41-year-old account executive feels "proud" to be a father (and hold a job) because he is able to spend more time with his sons than his father spent with him when he was a child.

> Being a father really makes me feel proud. Probably [because I] try to understand my sons more than my dad tried to understand me. Of course my dad worked all the time. He had his own business and he was hardly ever home. I think that is probably why [I feel proud because] I [spend more] time with them than my father [spent with me]. My father worked twelve, fourteen hours and sometimes sixteen hours a day. We very seldom saw him. Just maybe on Wednesday afternoons and Sundays. That was about it. I guess it goes back to the way we've been raised. (#1061M)

While the men in the previous two examples compared themselves to their own fathers, other men compared themselves to employed male peers such as their friends and coworkers. However, similar to the men in the previous examples, these men also emphasized how much more time they spend with their children relative to the men in their reference group. A 29-year-old aseptic operator, whose work schedule allows him to share child care with his wife, compared himself to other men at his stage of life who work a more usual eight-to-five day.

I take care of my kids the best I can. I'm actually with my kids a lot more than most fathers because I get up in the morning with them, feed them breakfast and lunch, and sometimes give them a bath. I spend anywhere from seven to eight [in the morning] until two o'clock [in the afternoon] with them. So that's six to seven hours a day I get to spend with my kids. A lot of fathers that work days get home at five o'clock and their kids are in bed by nine so [they're with their kids] four hours. I sometimes get twice as much time with my kids than my friends from work. And we do lots of things. (#1075M)

For men whose reference group consisted of male peers, social comparisons also were associated with positive self-evaluations as fathers and husbands because they were relatively more active in family life. This same respondent evaluated himself as a "pretty good father" and also projected what his coworkers *would* do with their children *if* they had his work hours.

I'm a pretty good [father] because I'm with them a lot. I really do take care of them. There's a lot [of] guys I know that would, if they had the same shift I have, would take their kids to day care at ten o'clock in the morning and then just have all that time to do nothing. I spend a lot of time, as much time with them as I can. (#1075M)

Later in the interview, this same respondent continued to describe how much more involved he is in home life compared to his own father and brother as well as the "other guys he knows."

I'm a lot more involved with my kids than a lot of other guys I know. I've got a few friends that don't do nothing. I mean they get along okay with their wives, but they just don't do nothing. They come home from work and hit the couch and that's where they stay most of the night. They don't help out with kids and stuff like I do. As a matter of fact, my brother and my dad are the same way. They don't help. They don't do nothing. (#1075M)

In addition to positive self-evaluations and feelings of pride, most of the men also indicated that combining employment with marriage and parenthood makes them feel satisfied and worthwhile. The next response from a 49-year-old teacher represents this theme.

[I feel] very satisfied and very worthwhile. I would think I was worthless if I didn't work. (#0163M)

When I probed this father's response and asked him why combining work and family roles contributed to his feelings of self-satisfaction and self-worth, he referred to the social expectations and norms he learned in childhood— which is that men *should* be employed.

> I was always expected to work. I guess I grew up in an environment where it was frowned upon if somebody didn't work. I grew up in a family where the women did not work outside of the home. My dad worked two jobs most of his life. I just grew up that way. I think that's a lot of it. (#0163M)

Overall, employment (and multiple role occupancy) resulted in positive self-evaluations for many men as fathers and husbands and positive feelings such as pride, self-satisfaction, and self-worth. The examples presented here suggest that the groups and individuals men selected as their frame of reference may have contributed to their positive self-appraisals and emotions. As was the case for women, in the absence of clear norms and objective standards for evaluating their role performances as husbands and fathers, men engaged in social comparisons with other *men*. When evaluating themselves as husbands and fathers, several men spontaneously compared themselves to their own fathers and other *employed* male peers, who they perceived as being less involved than they were in family life. By choosing as their reference group a more traditional model of noninvolved husbands, social comparisons led to perceptions of their own success as fathers and husbands, since they were relatively more active in family life. Although they spent far less time than their wives did in family life, by using other men as their frame of reference, these husbands perceived that they not only lived up to, but actually surpassed the standard of their reference group. In addition to having positive feelings of success and pride, these men also experienced feelings of self-satisfaction and self-worth from combining multiple roles because, in spite of their employment, they managed to be actively involved in family life.

Given the importance of family roles to men, it is understandable why the psychological benefits of employment and multiple role occupancy are greater for them relative to women. Insofar as family roles are salient to men, it is reasonable to expect that favorable self-evaluations as fathers and husbands have positive consequences for their emotional functioning. Positive self-evaluations and a configuration of positive emotions (including feelings of pride and self-satisfaction) in important role domains—which, in part, reflect the groups men select as their frame of reference—increase their feelings of self-worth, which in turn, may contribute to their relatively low symptom levels and relatively high levels of psychological well-being.

Men's Reference Groups: The Consequences of Having an Employed
Wife for Husbands' Negative Self-Evaluations and Feelings of Inadequacy

It is, however, important to mention that while employment (and multiple role occupancy) resulted in positive self-evaluations and positive feelings for the majority of men, there was a subgroup of men in the sample (N = 7) who felt inadequate as fathers and husbands because their wives were employed. The men who felt inadequate as husbands and fathers consisted of those who preferred that their wife not be employed and whose wife also preferred not to be employed (but the wife was employed primarily for financial reasons). According to these men, their wives' employment threatened their identities as a "good" husband and father because it *meant* that they were not able to adequately "provide." The emotions these men experienced were consistent with their evaluations of their performances in family role domains. In addition to feelings of inadequacy as providers, these men also experienced negative feelings of sadness, remorse, and guilt because their wives had to be employed to make ends meet at a time when their children were young.

When I probed their responses and asked them why having an employed wife resulted in negative self-evaluations and emotions, a few of these men (N = 4) responded by comparing themselves to other groups and individuals. The groups and individuals to which these men compared themselves tended to consist of other *employed* men, including their own fathers, male peers, and other childhood role models, who were *able* to provide a "family wage." By comparing themselves to an earlier generation of single wage-earner husbands, these men perceived that their family role behavior did not live up to the standard of their reference group.

For example, although combining work and family obligations makes this 31-year-old accounting clerk feel "happy," he nevertheless talked about his feelings of guilt and inadequacy as a provider.

> I'm really sad that she has to work and I'm really guilty that she has to work because I feel like I can't provide enough. I get that from, I think [I get that from] myself and I also think [I get that from] my parents. [My mother] didn't have to [work] and things like that. (#0256M)

When I probed and asked him why he feels sad and guilty that his wife is employed, he referred to his wife's employment preferences and emphasized the fact that he (and his wife) grew up in families in which the "mother could stay-at-home."

> She'd just rather be at home with the kids. That would've been her choice if we could've made it that way. And so that puts some stress on her. We both feel

guilty that the children have to go to day care and aren't at home because both of us were raised in families where the mother could stay at home. It all stems back to the way we were brought up. (#0256M)

In addition to feelings of inadequacy and guilt, a few of these men also felt remorse. For example, although this next 49-year-old computer programmer realized that "most families are two-income families," he nevertheless engaged in social comparisons with a male peer who he perceived as economically more successful than himself.

I don't feel good about it. [I feel] a little guilty that she has to bring in a paycheck to make ends meet. I wish I had a job good enough so she didn't have to [work]. I wish I made enough so she didn't have to work. [I feel] resignation. That's just the way it is and I'm not alone. Most families are two-income families. [I] regret that I didn't think about it earlier in life so that I'd be in a [different] financial position. The paper had an article about a fellow who graduated from the same high school I did four years before I did. He built a company from nothing. It's a two billion dollar a year business now. I didn't prepare myself for employment. (#0081M)

Whereas the men in the previous two examples compared themselves to the "type" of family they grew up in or financially more successful male peers, the next 40-year-old service technician compared himself to another childhood role model. While discussing his (and his wife's) preferences for her employment, I asked him why he prefers that his wife not work outside the home. He revealed his feelings of inadequacy as a provider when he compared himself to media images of a "traditional family" that he was exposed to as a child.

I've always thought [that in] the traditional family, the mother stays home. Like watching "Leave It To Beaver" or something. When Wards at work the mother's at home with Beaver and all that stuff. Maybe if things were different in my life. If I'd gone to college and maybe had a better job we wouldn't be in a situation that we're in. I think she carries a big load with the family responsibilities as well as her working responsibilities as well as being a wife and mother. Mothers don't have it easy. (#0072M)

In short, although employment and multiple role occupancy resulted in positive self-evaluations and positive emotions for most men, having a *wife* who combines multiple roles resulted in negative self-evaluations as husbands and fathers and negative feelings of inadequacy, sadness, remorse, and guilt for a subgroup of men (see Ross, Mirowsky, and Huber 1983). The examples discussed here suggest that the groups and individuals these men selected as their frame of reference may have contributed to their negative self-appraisals and emotions. In a climate in which men's and women's social roles are ostensibly undergoing redefinition and change, men (like women) appear to engage social comparisons with other people of the *same* sex. When evaluating themselves as fathers, husbands, and providers, this particular group of men compared themselves to their own fathers, male peers, and/or childhood role models who they perceived as more successful in work *and* family role domains. By choosing as their reference group an earlier generation of traditional single-wage earner husbands, social comparisons led to perceptions of their own inadequacy as fathers, husbands, and providers because they did not live up to the standard of their reference group. As in the case of the women, it appeared that these men's reference group served both as a standard for social comparisons as well as a source of norms regarding appropriate family role behavior for males. In addition to experiencing negative feelings of inadequacy, these men also felt guilty toward their children and wives because they were not able to live up to the "breadwinner" ideal which was embodied by their reference group.

Given the importance of family role identities for men's and women's self-conceptions, it is understandable why having an employed wife is both stressful and distressing for some husbands (Kessler and McRae 1982; Rosenfield 1980). To the extent that family role-identities (including the breadwinner identity) are important sources of self-definition for men, it is reasonable to assume that unfavorable self-evaluations as fathers, husbands, and breadwinners have negative consequences for their psychological well-being. In a parallel manner to the women discussed earlier, negative self-evaluations and a configuration of negative emotions (including feelings of inadequacy, sadness, remorse, and guilt) in these salient role domains—which, in part, reflect the groups and individuals these men select as their frame of reference—decrease their feelings of self-worth, which in turn, may contribute to their high symptom levels and poor mental health relative to other more "successful" husbands.

DISCUSSION: THE IMPORTANCE OF REFERENCE GROUPS AND SELF-EVALUATIONS FOR THEORY AND RESEARCH ON THE MENTAL HEALTH CONSEQUENCES OF MULTIPLE ROLES

Reference groups, and the social comparisons based on such groups, are important sources of information about the self. In order to interpret whether their role performances represent successes or failures, individuals routinely engage in social comparisons with either a group or an individual. In addition to providing comparative information, reference groups are also a source of normative information, since it is through these groups that individuals ascertain norms governing role-related behavior. Reference groups appear to be particularly important sources of comparative and normative information when objective standards and clear norms for role behavior are unavailable. Previous work on this topic indicates that because reference groups provide self-relevant information, the choice of a reference group has consequences for self-evaluations as well as feelings of inadequacy or self-worth. However, while several sociological theories have highlighted the contributions of reference groups to a broad range of social psychological phenomena (including the formation and maintenance of the self-concept), theories about the mental health effects of social roles have not considered the potential importance of social comparisons in the stress process.

In this paper, I have argued that reference groups, and the self-evaluations based on these groups, may be implicated in gender differences in mental health. Drawing on earlier conceptual and empirical work on reference group behavior, I suggested that the groups and individuals men and women select as their frame of reference provide further insight into why the psychological benefits of employment (and multiple role occupancy) are fewer for women relative to men. On the basis of previous theory and research on reference groups, social comparisons, the self-concept, and mental health (as well as insights from identity theory), it seemed reasonable to expect that favorable social comparisons result in positive self-evaluations, positive feelings of self-worth, as well as emotional well-being, especially if the role performance being evaluated is in a psychologically salient role domain. Conversely, unfavorable social comparisons should result in negative self-evaluations and negative feelings of inadequacy that are troublesome for mental health, particularly if the role-identity being evaluated is central to the individuals self-concept. To the extent that recent social change in men's and women's social roles and role configurations has produced ambiguity in the standards and norms for gender specific role behavior—at the same time that family roles have continued to be highly important sources of self-definition for both males and females— an examination of the groups men and women employ as their frame of reference may help us understand why combining work and family roles is both more stressful and distressing for women than for men.

To illustrate the potential importance of reference groups and social comparisons for men's and women's self-evaluations and emotional experience, I provided a few examples of reference group behavior from a qualitative study of gender differences in the emotional effects of multiple role involvements. Although I did not originally intend to examine men's and women's reference groups, data from in-depth interviews with employed married parents revealed that differences between men's and women's evaluations of themselves as spouses and parents, and their feelings about combining work and family roles, were associated with the different groups they employed as their frame of reference. It is likely that differences between men's and women's evaluations of themselves as spouses and parents, and their feelings about combining work and family roles, are precipitating factors in the production of gender differences in psychological symptoms such as depression, anxiety, somatization, and distress.

On the basis of these examples, it was evident that when evaluating themselves as spouses and parents, employed married mothers tended to compare themselves to their own mothers, or other nonemployed wives, who were more involved than they were in family life. Social comparisons were associated with negative self-evaluations (as wives and mothers) and negative feelings of inadequacy and self-doubt for employed mothers because their role behavior fell short of the standard of their reference group. In contrast, when assessing themselves as spouses and parents, employed married fathers tended to compare themselves to their own fathers, or employed male peers, who were less involved than they were in family life. Social comparisons were associated with positive self-evaluations (as husbands and fathers) and positive feelings of success, pride, self-satisfaction, and self-worth for employed husbands because their role behavior surpassed the standard of their reference group.

However, while multiple role occupancy resulted in positive self-evaluations and feelings of self-worth for most husbands, I also found that having an employed wife resulted in negative self-evaluations and feelings of inadequacy for a subgroup of men. When assessing themselves as husbands, fathers, and breadwinners, the men who preferred that their wives not be employed and whose wives also preferred not to be employed (but the wife was employed primarily for financial reasons) tended to compare themselves to their own fathers and other role models of single wage-earner husbands whose wives were able to stay at home with their children. Social comparisons were associated with negative self-evaluations (as husbands and fathers) and negative feelings of inadequacy and remorse for these men because their role behavior did not live up to the standard of their reference group.

These examples also suggested that the groups and individuals men and women employed as their frame of reference served not only as a standard of behavior, but also as a source of norms about gender specific role behavior and how they ought to act as spouses and parents. In addition to negative self-evaluations and feelings of inadequacy, social comparisons were associated with negative feelings of guilt for most women and for some men because their role behavior violated the norms and expectations of gender appropriate role behavior, which were embodied by their respective reference groups.[8] Not surprisingly, gender variation in self-evaluations, feelings of self-worth or inadequacy, and positive as opposed to negative affective states, which were evident in this small sample, parallel distress differences between women and men (and among men) that are consistently reported in both community and national epidemiological studies (e.g., Kessler and McRae 1982; Menaghan 1989; Rosenfield 1980; Thoits 1986).

Given the small nonrepresentative sample and method of analysis, the findings and conclusions discussed in this paper are obviously only suggestive and tentative. In order to determine whether men's and women's evaluations of themselves as spouses and parents (and their feelings about combining work and family roles) are additional mediating variables in the gender-distress relationship, the ideas I have outlined would have to be subjected to rigorous "tests" with longitudinal quantitative data from men and women from a range of class, ethnic, and racial backgrounds. Moreover, due to data limitations, I was unable to examine the reference group selection process and, therefore, cannot rule out the alternative hypothesis that men and women select reference groups that sustain preexisting self-evaluations and feelings. Panel studies are necessary to examine the selection process and to determine the causal direction of the relationship between reference groups and self-evaluations. However, while my data did not allow me to explain *why* men and women selected these reference groups for social comparisons, by showing that employed married mothers' and fathers' reference groups consisted of *same sex* individuals, usually *family members* and/or *peers*, who themselves were spouses and parents, my findings are consistent with social comparison theory's claim that people choose referents who are, in some way, similar or close to themselves (Festinger 1954).

By providing further insight into the social psychological processes that may underlie distress differences between men and women who have the same role configuration and role situation, the examples presented earlier help identify some new questions for future surveys on gender, social roles, and mental health. For example, in addition to the need to develop varied measures of self-assessed role performance across different role domains, surveys should ask men and women to describe the feelings they experience from combining work and family obligations as well as their feelings about having an employed spouse. Surveys also should obtain information about the groups (and/or individuals) men and women employ for social comparisons, and if social comparisons result in positive or negative self-appraisals and feelings. To further elaborate the sources of gender differences in self-evaluations and emotional experience, surveys also should include questions that assess individuals' reflected appraisals. Although I have focused on reference groups and social comparisons, reflected appraisals are also an important source of evaluative information about the self. Therefore, in addition to the questions mentioned previously, men and women should be asked how others (including their spouse, children, parents, friends, neighbors, coworkers, and employers) perceive them as spouses, parents, and workers. It is likely that theory and research on gender, social roles, and mental health would be enhanced by answers to these questions. However, whether the inclusion of these variables in statistical models would reduce gender differences in psychological symptoms is not resolvable at this time and awaits further research.

Beyond highlighting the relationships between reference groups, self-evaluations, and emotional experience among individuals who hold the same three roles of spouse, parent, and worker, the ideas discussed in this paper may also be useful for theory and research about the mental health consequences of social roles more generally. The findings reported in this study suggest that self, identity, and emotion processes are all implicated in the etiology of psychological well-being and emotional distress. Epidemiological studies that focus on social status differences in mental health would surely be enhanced by systematic attention to these underlying social psychological processes. Moreover, by linking the social distribution of well-being and distress to the social distribution of positive and negative self-evaluations and affect in the general population, such a research program also would enhance sociological theory and research on self, identity, and emotion.

Finally, by identifying the groups men and women employ for social comparisons, the findings discussed in this paper provide a window into why recent social change in men and women's roles has not reduced the gender gap in mental health among married persons. While the increase in women's employment has resulted in a convergence of the role configurations and role situations of males and females, the groups men and women select as the basis for self-evaluation nevertheless appear to consist of role models of traditional wives and husbands to which they were exposed in childhood. To the extent that men's and women's reference groups represent and embody standards and norms of traditional gendered family role behavior, it is not surprising that employment (and multiple role occupancy) result in feelings of inadequacy and distress for women, and feelings of self-worth and well-being for men. If the dual-earner family continues to be a predominant family form, we could perhaps expect that future cohorts of employed married parents will select reference groups consisting of males and females who are actively involved in *both* work *and* family life. With greater availability of alternative reference groups, we may in the future see a decline in current discrepancies between men and women in the psychological benefits of multiple role occupancy. In short, by identifying gender differences in reference groups, self-evaluations, and emotional experience among employed married parents, the ideas and findings discussed in this paper provide insight into one mechanism which

contributes to the persistence of gender differences in mental health among
married persons.

APPENDIX

Selected Sociodemographic Characteristics
of the Follow-Up Sample by Gender

	Follow-Up Sample		
	Total (N = 40)	Male (N = 20)	Female (N = 20)
Characteristics			
Age, Mean Years	38.0	40.0	36.0
25-34	35.0%	25.0%	45.0%
35-44	42.5%	45.0%	40.0%
45-54	20.0%	25.0%	15.0%
55-64	2.5%	5.0%	0%
Race			
White	100%	100%	100%
Black	0%	0%	0%
Other	0%	0%	0%
Education			
Less Than High School	0%	0%	0%
High School Graduate	25.0%	25.0%	25.0%
Some College	45.0%	45.0%	45.0%
College Graduate	15.0%	15.0%	15.0%
Graduate Degree	15.0%	15.0%	15.0%
Household Income			
Under $12,000	0%	0%	0%
$12,000-19,999	0%	0%	0%
$20,000-31,999	7.5%	5.0%	10.0%
$32,000-39,999	15.0%	15.0%	15.0%
$40,000-51,999	40.0%	40.0%	40.0%
$52,000-59,999	15.0%	10.0%	20.0%
$60,000-71,999	10.0%	10.0%	10.0%
$72,000 or more	12.5%	20.0%	5.0%
Children < 18 years Residing in the Household, Mean No.	2.0	2.1	1.9
Employment			
Employed 35hrs +/wk	100%	100%	100%
Spouse Employed 35hrs +/wk	100%	100%	100%

ACKNOWLEDGMENT

I am grateful to Karen Heimer, Jodi O'Brien, and especially Brian Powell for their detailed comments on an earlier version of this paper. Peggy Thoits and Sheldon Stryker provided helpful suggestions on the larger research project upon which this manuscript is based. I appreciate the careful research assistance of Nicholas Pedriana who spent long hours coding data. I also thank the National Institute of Mental Health and The University of Iowa for financial support through a National Research Service Award (No. MH-10110) and two Old Gold Summer Fellowships.

NOTES

1. Shibutani (1955) noted that while the concept of the reference group provides additional refinements to some tenets of symbolic interactionism, the processes involved in reference group behavior are not fundamentally different from the socialization process previously discussed by symbolic interactionists such as Mead (1934). From a symbolic interactionist perspective, one can think of reference *groups* and reference *individuals* as "particular others" whose perspectives are considered in the role-taking process.

2. In addition to having consequences for individuals, reference groups also have consequences for processes involving group formation and maintenance. However, because this paper is concerned with the effects of social comparisons for *self* and *emotion* processes, I have not included a discussion of the implications of reference group behavior for group processes.

3. See Tesser, Millar, and Moore (1988) for an example of psychological research that examines the consequences of social comparisons for self-evaluations of performances on "self-relevant" tasks.

4. While Hyman (1942) is credited for having first introduced the term "reference group," as noted previously, early symbolic interactionists such as Mead (1934) had previously acknowledged the importance of reference groups for self-development.

5. Psychological symptoms assessed in epidemiological studies are typically measured by standard screening scales such as the Center for Epidemiological Studies' Depression Scale (CES-D) and the SCL-90 (Derogatis and Cleary 1977). These screening scales are comprised of items which were identified by the presenting complaints of patients receiving psychiatric treatment. When combined, these items measure psychological symptoms such as depression, anxiety, somatization, and distress. These scales have been shown to have high construct validity and high internal consistency in general population surveys. While not measures of psychiatric disorder per se, scores on these measures are interpreted as reasonably good indicators of experienced psychological distress. In this paper, I use the terms psychological well-being, distress, mental health, and emotional functioning interchangeably; consistent with their usage in the mental health literature, I use these terms to refer to individuals' *psychological state*. The term "stress" is used to refer to a *state of arousal* resulting from demands from the environment that are perceived by the individual as taxing. Sociological stress researchers generally regard "stress" as a mediating variable in the relationship between social circumstances (e.g., social roles or multiple role occupancy) and psychological outcomes such as depressive symptomatology (see Aneshensel [1992] for a recent review of theory and research on social stress).

6. Although this respondent no longer felt guilty holding a job now that her children were older, other women who had older children experienced negative feelings from combining work and parenthood.

7. All names in quotes are pseudonyms. The following notations are used in examples from transcripts: [ ] words in brackets refer to an inaudible response which I surmised based on interview notes and/or the context of the conversation and...refers to a response that trails off and was not completed by the speaker.

8. Although my research suggests that men and women employ different reference groups for social comparisons and self-evaluations, men's and women's reference groups nevertheless appear to have three things in common; they are comprised of *same sex* persons, usually *childhood role models* of spouses and parents, who represent *traditional* family role obligations of males and females.

REFERENCES

Aneshensel, C. S., R. R. Frericks, and V. A. Clark. 1981. "Family Roles and Sex Differences in Depression." *Journal of Health and Social Behavior* 22: 379-393.

Aneshensel, C. S. 1992. "Social Stress: Theory and Research." *American Review of Sociology* 18: 15-38.

Cleary P. and D. Mechanic. 1983. "Sex Differences in Psychological Distress Among Married People." *Journal of Health and Social Behavior* 24: 111-121.

Crosby, F. 1976. "A Model of Egoistical Relative Deprivation." *Psychological Review* 83: 85-113.

Derogatis, L. and P. Cleary. 1977. "Confirmation of the Dimension Structure of the SCL-90: A Study of Construct Validation." *Journal of Clinical Psychology* 33: 981-989.

Easterlin, R. A. 1973. "Does Money Buy Happiness?" *The Public Interest* 30: 3-10.

Elder, G. H., Jr. 1974. *Children of the Great Depression.* Chicago, IL: University of Chicago Press.

Eisenstadt, S. N. 1968. "Studies in Reference Group Behavior." Pp. 413-429 in *Readings in Reference Group Theory and Research*, edited by H. Hyman and E. Singer. New York: Free Press.

Felson, R. B. and M. D. Reed. 1986. "Reference Groups and Self-appraisals of Academic Ability and Performance." *Social Psychology Quarterly* 49: 103-109.

Festinger, L. 1954. "A Theory of Social Comparison Processes." *Human Relations* 7: 117-140.

Form, W. H. and J. A. Geshwender. 1962. "Social Reference as a Basis of Job Satisfaction: The Case of Manual Workers." *American Sociological Review* 27: 228-237.

Gore, S. and T. W. Mangione. 1983. "Social Roles and Psychological Distress: Additive and Interactive Models." *Journal of Health and Social Behavior* 24: 300-312.

Gove, W. and M. Geerken. 1977. "The Effects of Children and Employment on the Mental Health of Men and Women." *Social Forces* 56: 66-76.

Haw, M. A. 1982. "Women, Work, and Stress: A Review and Agenda for Future Research." *Journal of Health and Social Behavior* 23: 132-144.

Hyman, H. H. 1942. "The Psychology of Status." *Archives of Psychology* 269.

————. 1960. "Reflections on Reference Groups." *Public Opinion Quarterly* 24: 383-396.

Hyman, H. H. and E. Singer. 1968. *Readings in Reference Group Theory and Research.* New York: Free Press.

Kandel, D. B., M. Davies, and V. H. Ravies. 1985. "The Stressfulness of Daily Social Roles for Women: Marital, Occupational, and Household Roles." *Journal of Health and Social Behavior* 26: 64-78.

Kasl, S. V. 1989. "An Epidemiological Perspective on the Role of Control in Health." Pp. 161-189 in *Job Control and Worker Health*, edited by S. L. Slaute, J. J. Hrrell, and C. L. Cooper. London: Wiley.

Kelley, H. H. 1968. "Two Functions of Reference Groups." Pp. 77-83 in *Readings in Reference Group Theory and Research*, edited by H. Hyman and E. Singer. New York: Free Press.

Kessler, R. C. and J. A. McRae, Jr. 1982. "The Effect of Wives' Employment on the Mental Health of Married Men and Women." *American Sociological Review* 47: 217-227.

Lennon, M. C. and S. Rosenfield. 1992. "Women and Mental Health: The Interaction of Job and Family Conditions." *Journal of Health and Social Behavior* 4: 316-327.

Link, B., M. C. Lennon, and B. P. Dohrenwend. 1993. "Socioeconomic Status and Depression: The Role of Occupations Involving Direction, Control, and Planning." *American Journal of Sociology* 98: 1351-1387.

Loscocco, K. A. and G. Spitze. 1990. "Working Conditions, Social Support, and the Well-being of Male and Female Factory Workers." *Journal of Health and Social Behavior* 31: 313-327.

Lowe, G. S. and H. C. Northcott. 1988. "The Impact of Working Conditions, Social Roles, and Personal Characteristics on Gender Differences in Distress." *Work and Occupations* 15: 55-77.

Mead, G. H. 1934. *Mind, Self, and Society*. Chicago: The University of Chicago Press.

Menaghan, E. G. 1989. "Role Changes and Psychological Well-being: Variations in Effects by Gender and Role Repertoire." *Social Forces* 67: 693-714.

Merton, R. K. and A. S. Rossi. 1950. "Contributions to the Theory of Reference Group Behavior." Pp. 40-105 in *Continuities in Social Research: Studies in the Scope and Method of the American Soldier*, edited by R. K. Merton and P. F. Lazarsfield. Glencoe, IL: Free Press.

Miller, J., C. Schooler, M. Kohn, and K. Miller. 1979. "Women and Work: The Psychological Effects of Occupational Conditions." *American Journal of Sociology* 85: 66-94.

Newcomb, T. M. 1943. *Personality and Social Change*. New York: Holt, Rinehart, and Winston.

Parker, S. and R. J. Kleiner. 1968. "Reference Group Behavior and Mental Disorder." Pp. 350-373 in *Readings in Reference Group Theory and Research*, edited by H. Hyman and E. Singer. New York: Free Press.

Patchen, M. 1961. "A Conceptual Framework and Some Empirical Data Regarding Comparisons of Social Rewards." *Sociometry* 24: 136-156.

Pearlin, L. 1975. "Sex Roles and Depression." Pp. 191-207 in *Life Span Developmental Psychology: Normative Life Crises*, edited by N. Datan and L. Ginsberg. New York: Academic Press.

Pettigrew, T. 1968. "Actual Gains and Psychological Losses." Pp. 339-349 in *Readings in Reference Group Theory and Research*, edited by H. Hyman and E. Singer. Free Press: New York.

Radloff, L. 1975. "Sex Differences in Depression: The Effects of Occupation and Marital Status." *Sex Roles* 1: 249-265.

Roberts, R. E. and S. J. O'Keefe. 1981. "Sex Differences in Depression Reexamined." *Journal of Health and Social Behavior* 22: 394-400.

Rosenberg, M. 1979. *Conceiving the Self*. New York: Basic Books.

Rosenberg, M. and L. Pearlin. 1978. "Social Class and Self-esteem Among Children and Adults." *American Journal of Sociology* 84: 53-77.

Rosenberg, M. and R. Simmons. 1972. *Black and White Self-Esteem: The Urban School Child*. Caroline Rose Monograph Series. Washington, DC: American Sociological Association.

Rosenfield, S. 1980. "Sex Differences in Depression: Do Women Always Have Higher Rates?" *Journal of Health and Social Behavior* 21: 33-42.

_____ . 1989. "The Effects of Women's Employment: Personal Control and Sex Differences in Mental Health." *Journal of Health and Social Behavior* 30: 77-91.

_____ . 1992. "The Costs of Sharing: Wives' Employment and Husbands' Mental Health." *Journal of Health and Social Behavior* 33: 213-225.

Rosow, I. 1967. *Social Integration of the Aged.* New York: Free Press.

Ross, C., J. Mirowsky, and J. Huber. 1983. "Dividing Work, Sharing Work, and In-between: Marital Patterns and Depression." *American Sociological Review* 48: 809-823.

Shibutani, T. 1955. "Reference Groups as Perspectives." *American Journal of Sociology* 60: 562-569.

Simon, R. W. 1992a. "Parental Role Strains, Salience of Parental Identity, and Gender Differences in Psychological Distress." *Journal of Health and Social Behavior* 33: 25-35.

————. 1992b. "Spouse, Parent, and Worker: Gender, Multiple Role Involvements, Role Meaning, and Mental Health." Unpublished Doctoral Dissertation. Indiana University, Bloomington, IN.

————. 1995. "Gender, Multiple Roles, Role Meaning, and Mental Health." *Journal of Health and Social Behavior.* Forthcoming.

Singer, E. 1981. "Reference Groups and Social Evaluations." Pp. 66-93 in *Social Psychology: Sociological Perspectives,* edited by M. Rosenberg and R. Turner. New York: Basic Books.

Stern, E. and S. Keller. 1953. "Spontaneous Reference Groups in France." *Public Opinion Quarterly* 17: 208-217.

Strauss, H. M. 1968. "Reference Group and Social Comparison Processes Among the Totally Blind." Pp. 222-237 in *Readings in Reference Group Theory and Research,* edited by H. Hyman and E. Singer. New York: The Free Press.

Stryker, S. and A. Statham. 1985. "Symbolic Interactionism and Role Theory." Pp. 311-378 in *The Handbook of Social Psychology* (3rd edition), edited by G. Lindzey and E. Aronson. New York: Random House.

Suls, J. M. and R. L. Miller. 1977. *Social Comparison Processes.* New York: Wiley.

Tesser, A., M. Millar, and J. Moore. 1988. "Some Affective Consequences of Social Comparison and Reflection Processes: The Pain and Pleasure of Being Close." *Journal of Personality and Social Psychology* 54: 49-61.

Thoits, P. A. 1986. "Multiple Identities: Examining Gender and Marital Status Differences in Distress." *American Sociological Review* 51: 259-572.

————. 1991. "On Merging Identity Theory and Stress Research." *Social Psychology Quarterly* 54: 101-113.

————. 1992. "Identity Structures and Psychological Well-being: Gender and Marital Status Comparisons." *Social Psychology Quarterly* 55: 236-256.

————. 1995. "Identity-Relevant Events and Psychological Symptoms: A Cautionary Tale. *Journal of Health and Social Behavior.* Forthcoming.

Turner, R. H. 1956. "Role Taking, Role Standpoint, and Reference Group Behavior." *American Journal of Sociology* 61: 316-328.

Walster, E., G. W. Walster, and E. Berscheid. 1978. *Equity: Theory and Research.* London: Allyn and Bacon.

MODELING INDIVIDUAL PERCEPTIONS OF JUSTICE IN STRATIFICATION SYSTEMS

Yuriko Saito

ABSTRACT

Fararo (1973) formulated a model of how individuals form mental images of social class systems. Building from that model, this paper constructs an axiomatic theory to explain how people judge the fairness of their society. The basic idea is that these judgments are based on a notion of "who gets what" in the society. From previous empirical findings, a meritocratic allocation system is postulated to exist. First, I propose axioms concerning how actors obtain an image of allocation in a meritocratic society through their daily interaction with others. Two types of information-seeking about others are specifically proposed and discussed. In either case, axioms imply that actors arrive at stable images that vary with their locations in the allocation system. Second, I state axioms for a fairness judgment process based on the image. These axioms imply that actors in various locations unanimously reach a single conception of "fair society" from their images, if the allocation system had equally differentiated systems of merit and reward. These axioms also suggest that no one feels society is completely fair.

Advances in Group Processes, Volume 12, pages 51-80.
Copyright © 1995 by JAI Press Inc.
All rights of reproduction in any form reserved.
ISBN: 1-55938-872-2

Perception of fairness has been one of the central issues in social psychology for nearly three decades. We now have several theories of fairness judgments, in addition to a large body of empirical findings. Among the theories are status-value and reward expectation theories (Berger, Zelditch, Anderson, and Cohen 1972; Berger, Fisek, Norman, and Wagner 1983), Jasso's (1980) distributive justice theory, and Markovsky's (1985) multilevel justice theory. The allocation of goods is at a relatively micro-level in these theories: all deal with fairness in small groups. By "small" I mean group sizes that make it possible for an individual to have precise, first-hand knowledge regarding "who gets what."

Fairness evaluations of larger-scale objects such as societies have not been a central focus in previous justice theories. However, issues of fairness or justice were essential for ancient philosophers such as Plato and Aristotle, particularly because they acknowledged its immense communal importance—an issue still pertinent today. In particular, there is always a possibility, perhaps greater now than in ancient times, of large-scale change if enough people view the current status quo as illegitimate. In this regard, the fairness evaluation of a society is a problem worth pursuing practically as well as theoretically.

To explain perceptions of societal fairness, something must be added to the earlier theories. This is because, unlike the case in small groups, "who gets what" in a society cannot be directly perceived. Thus, we first need to examine the process by which an actor acquires a notion of "who gets what." Saito (1987) presented a model to describe this process. It combined an "image formation" model (Fararo 1973) and a theory of justice evaluations. This paper develops and extends the model within the same framework. Since the model presented in Saito (1987) forms the basis of discussion, it will be introduced first.

The basic premise in the model constructed in this chapter is that the idea of "who gets what" is generated from the actor's image of the allocation system and, consequently, the fairness evaluation depends on that image. The model takes the form of an axiomatic theory. The axioms are largely divided into two categories, one dealing with the process of image formation and the other with fairness judgments based on the image. In order to evaluate the fairness of an allocation, an actor needs to know "what is fair for whom" as well as "who gets what." Accordingly, questions to be addressed in this paper are: (1) How does the image of an allocation system depend on the actual allocation system and on the position of the actor? (2) What does the actor regard as fair for a person in a given position? (3) How (un)fair is the society from the actor's perspective?

Before starting the formal discussion, I will briefly review the relevant empirical findings, limited as they are. First, the meritocratic allocation system, which appears to operate in current capitalist societies, is regarded as legitimate by the majority of Americans (Robinson and Bell 1978; Robinson 1983; Kluegel and Smith 1981). This belief in meritocracy is not affected by individually experienced economic problems (Kluegel 1988). These findings suggest that an equity rule (that allocations based on contribution or human capital investment are fair) is regarded as a fair allocation principle by Americans, and possibly by other members of the capitalist world.

Second, people can quite accurately estimate the income of a person based on his or her social characteristics, although the impact of ascribed characteristics is underestimated (Sheplak and Alwin 1986). From this we can infer that people may "survey" the social characteristics and rewards of others whom they contact.

Third, the results of this survey may be systematically distorted due to the social distance between the actor and the others. Although respondents in all occupations more or less misperceive the income of the top and bottom occupations, Headey (1991) found that the degree of errors is, to some degree, a function of the actor's distance from the evaluated occupation in the occupational status hierarchy. Thus, we can further infer that the "survey" may be conducted more thoroughly on others who are socially close.

Based on this empirical evidence, we can draw a rough sketch of how people form a mental image of the allocation system. First, the image is generated through ongoing surveys conducted by the actors themselves. Second, it is affected by the distance between the actor's and others' positions. Third, from Sheplak and Alwin's findings, actors must have prior conceptions of the social characteristics of others. Without such conceptions, they could not estimate income based on social characteristics. These elements will be integrated into the axioms presented in the next section.

The model presented here assumes the existence of an allocation system that assigns various amounts of reward (or outcomes) to actors with achieved and ascribed status characteristics (or inputs). Actors are assumed to acquire their ideas about "who gets what" by forming an image of this allocation system through their interactions with others. Specifically, during interactions they are assumed to engage in an information-seeking process. Actors are assumed to first probe the inputs of the other. They are less motivated to seek information about the outcomes of others with inputs that differ from their own. Thus, information-seeking is not carried out to the full extent for those whose inputs differ. The fairness judgment vis-à-vis the entire society is assumed to be based on this image of the allocation system. Utilizing their images, actors obtain a concept of fair outcomes for each input state, and compare it with the "actual reward" held in the image.

THE IMAGE FORMATION PROCESS
OF AN ALLOCATION SYSTEM

The first question to be answered in this paper is how the image of an allocation system depends on its actual form and the actor's social position. This section deals with the elementary image-formation process. The axioms in the first part of this section—as revision of Saito's (1987)—describe a process that produces an image with multiple equilibria not uniquely determined by the actor's position. The second section introduces an alternative premise about information-seeking. We will see that the image has an equilibrium totally dependent on the actor's position, if the nature of information-seeking is altered.

Model One

First, let us consider the objective allocation system that is the basis of the actor's image. In the previous section, empirical evidence suggested that Americans strongly endorse meritocracy. It is safe to say that meritocracies actually dominate resource allocation in the capitalist world. Meritocratic allocation consists of two elements: merit and reward. We can then reasonably postulate the existence of an allocation system that is described in Axiom A-1. Before stating it, let me introduce the basic terms.

In this model, I use the word "input" instead of "merit." It includes a person's ascribed and achieved characteristics. In this way, the model is potentially applicable to non-meritocratic allocation. Adams (1965, p. 277) originally defined input as "what a man perceives as his contributions to the exchange, for which he expects a just return." However, this concept was sometimes interpreted more broadly, placing more emphasis on "for which he expects a just return." I will use the concept in this broader sense, considering it as the allocation base in a given society.

Definition 1. Input and Outcome *Input* is a person's characteristic or contribution for which it is considered in the society that he or she can expect a return. *Outcome* is that which is considered to be a reward in a society.

Note that according to this definition, input does not need to be earned.

Definition 2: Rank *Ranks* are ordered, differentially evaluated states of inputs or outcomes.

Inputs and outcomes can be captured as sets with ranks as elements, for example, input ranks $I = \{1, 2, \ldots k, \ldots n\}$; outcome ranks $O = \{1, 2, \ldots k, \ldots n\}$, where the k'th element of I or O is denoted by i_k or o_k, respectively, and higher-numbered ranks indicate more highly-valued elements. Note that ranks are defined as a "consequence of evaluation" in the above definition. "High income" is not something just simply known; an actor evaluates a certain level of income as "high." Thus, this definition implicitly assumes comparisons between outcome states or input states by the actor. It is further assumed that enough agreement exists among actors on how to evaluate objects so that a uniform evaluation of input and outcome states becomes possible. Jackman (1979) suggested that there actually appears to be such uniform evaluation of input states. An allocation system, or a stratification system, can be considered as a system of such uniform evaluations.

Definition 3: Location of actor in allocation system   Actor X's input and outcome are denoted by I(X) and O(X), which are elements of the sets I and O, respectively. X's (objective) location in the allocation system is (I(X), O(X)).

Definition 4: Input Stratum   The set of actors whose input rank is k constitutes input stratum k or, simply, stratum k.

Next we turn to the axioms. Those numbered A-1 to A-9 are quoted from Fararo (1973), except revised axioms which are preceded by "*." Note that the theory's axioms constitute its main logical components. Later I will present several "assumptions," auxiliary theoretical components that are not part of the logical core.

* Axiom A-1:   There exists an allocation system S over a set A of actors in a time-domain T such that (1) S contains inputs and outcomes, each of which is a linearly ordered set representing a uniform, differential evaluation in A^1; (2) there is an ordering between the sets such that set I has priority over set O; (3) Based on (1) and (2), the set of all possible positions in S, which are the pairs of elements of I and O, is ordered "lexicographically" as illustrated in Figure 1.

In Figure 1 there are high and low input actors who receive high as well as low outcomes. Imagine a society allocating rewards based primarily on education level. Although the majority of the educated should be assigned a higher reward in this society, there would always be exceptions, for example, the educated receiving a lower reward. Two outcome levels corresponding to

Input   Outcome

High	
High	High
	Low
Low	High
	Low

Figure 1. A Lexicographical Allocation System

"high" input expresses this variation. Fairness in this system is defined as follows.

Definition 5: State of fairness
O(P) = O(GO), where I(P) = I(GO), and
I(A): input of actor A
O(A): outcome of actor A
P: actor whose state is the object of fairness judgment
GO: Generalized Other who is the referent of P

For the purpose of simplification, I assume there are the same number of input and outcome ranks in the allocation system:

Assumption 1: The number of ranks of the sets I and O are equal.

This means that degrees of differentiation in an allocation base (input) and resources to be allocated are assumed to be in perfect correspondence. If division of labor led people to make five evaluative distinctions about jobs, for example, then exactly five reward levels will be prepared in that society.

Let us assume this allocation system is a stable one, and there are people who do not experience any social mobility. Their image develops as they interact with others.

Axiom A-2: S is stable over T.

Axiom A-3: There exists a subset of actors in A such that their positions in S are time-invariant. Any such actor may be chosen for the role of focal actor in the analysis, that is, as the actor whose image is developing over time.

Axiom A-4: The focal actor has a series of interacts with members of A in time domain T. Denote points of T, in which such interacts occur for the focal actor, by t = 0, 1,.... At t = 0 the actor has an *initial* image of S.

The probability of encounters with alters in various positions should depend on their social distances. If there are N positions in the system, we obtain an N x N matrix of the probability of interaction. If the matrix contains a zero entry, there is no possibility for two given positions to interact. However, here I assume this is not the case: However small the probability, actors will eventually meet others in every position in the society.

Axiom A-5: Given an interact of a focal actor and some alter, ,the probability that the alter is in a certain position in S is not time-dependent; it depends only on the positions of the focal actor and alter.

Assumption 2: The probability of the focal actor's interaction with the alter in any position in S is assumed to be greater than zero.

Next, I offer the following premise about information-seeking:

*Axiom A-6: When a focal actor X interacts with alter Y, a process of information-seeking occurs as follows: First, X seeks information on Y's input. If Y's input belongs to the same rank as X's, X next seeks information about Y's outcome. If Y's input (say, i_k) differs from X's, X proceeds as follows: (1) if Y is the first encounter among the members of input stratum k, X seeks information on Y's outcome; (2) if Y is not X's first encounter, X's information-seeking stops when X gets information on Y's input. For Y's outcome rank, X substitutes the outcome rank of X's first encounter with Y's input stratum (see Figure 2).

As indicated in the latter part of Axiom A-6, the image of an input stratum that differs from X's is formed by generalizing first encounters. If the input

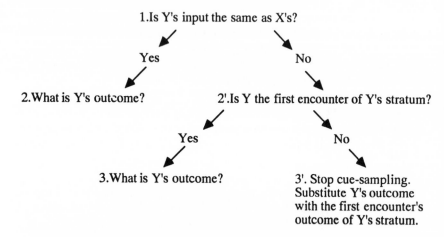

Figure 2.   Information Seeking Process by a Focal Actor: Model 1

rank differs from one's own rank, motivation for further information-seeking exists only when the focal actor lacks information about the outcome state of the input stratum.

Axiom A-6 claims that to judge whether or not society is fair, actors need to know something about the outcome of every stratum. At the same time, the axiom implies that the motivation of the actor to obtain outcome information about other input strata is not as strong as the motivation to know the outcome of the people with the same input level. This lesser motivation to know the outcome of different people has been empirically demonstrated (Coleman and Rainwater 1978). Searching for information only at the first encounter is a "compromise" of these two tendencies, that is, the focal actor wants to know just enough about other strata to form some image of their outcomes.

The claim of Axiom A-6 that first encounters with different strata have large impacts is justified by the findings on primacy effects in impression-formation literatures. But at the same time, there is a recency effect, that is, recent stimuli also have a large impact on image formation. I excluded this possibility from the model because the cost of information-seeking cannot be reduced if we hypothesize that the most recent encounter determines the image of different input strata. That the most recent encounter is most influential implies that the focal actor has to always seek information about outcomes. For the same reason, the obvious possibility of averaging outcomes of the alters from a given input stratum is excluded.

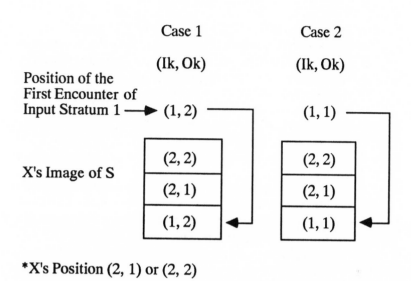

*X's Position (2, 1) or (2, 2)

Figure 3. Difference in Focal Actor X's Image of Allocation System due to First Encounter's Postion

Figure 3 shows how images depend on the location of the first encounter in a 2x2 system. In case 1, the focal actor X met (1, 2) first, thus his or her image of the input stratum 1 is (1, 2). Contrarily, in case 2, X met (1, 1) first, and this leads his or her image of the input stratum 1 to (1, 1). Since interaction is the only source of information for the actors, except for their own positions, the image should not be changed without it.

* Axiom A-7-1: Images change only as a result of acquiring information about S.

* Axiom A-7-2: A focal actor obtains information about S either from his or her own position in S or from interaction with others.

Naturally, actors do not know anything about others, but only about themselves, until they initiate interaction.

Axiom A-8: At t = 0, the image of S is the focal actor's position in S, (I(X), O(X)).

*Axiom A-9: Given an interact, the focal actor X's image is transformed according to the following: $P_X(Y)$ is Y's position as perceived by X. If Y's input differs from X's, and if Y is not the first encounter in Y's input stratum, then information-seeking stops after the input information is obtained, and Y's outcome is replaced by the outcome of the first encounter in Y's input stratum. This yields $P_X(Y) = (I(Y),$ O(first encounter)). Otherwise, $P_X(Y) = (I(Y), O(Y))$. Furthermore:

1. if $P_X(Y)$ is already included in the X's image of S, no change will occur in X's image.
2. if $P_X(Y)$ is located above the present top stratum of X's image, $P_X(Y)$ will become the new top stratum.
3. if $P_X(Y)$ is located below the present bottom stratum of X's image, $P_X(Y)$ will become the new bottom stratum.
4. if $P_X(Y)$ is located between two strata in the X's image, $P_X(Y)$ will be inserted between them.

Figure 4 illustrates the development of the image of actor X, whose position in S is (2, 1). She or he initially has her or his own position, (2, 1), as the image of S. Then, after interacting with the alter whose position in S is (2, 2), for example, X's image is transformed from (2, 1) to (2, 2), (2, 1). If instead X interacts with an alter whose position equals that of X's, then X's image of S does not change. Furthermore, in Figure 4, the position of the first encounter of input stratum 1 is assumed to be (1, 2). Note that according to Axiom A-9, the outcome information of the first encounter is used in the later interaction. Thus, an encounter with (1, 1) does not change X's image after X met (1, 2) (see Figure 4 panel (c) and (d)). Obviously, X's image would not change after it reaches the state "(2, 2), (2, 1), (1, 2)," because further encounters with alters of any position in S do not provide additional information. In this sense, X's image of S reaches an equilibrium.[2] Note that X in Figure 4 could have reached another equilibrium image, (2,2), (2,1), (1,1), if his or her first encounter in input stratum 1 were (1, 1) instead of (1, 2). Hence, X's image of S has two possible equilibria (stable images), depending on the position of the first encounter.[3] Figure 5 summarizes the final images of S, which has two ranks.

Figure 5 shows that the image of a focal actor in a given position reaches an equilibrium whose form is determined by the positions of the actor and first encounters. The place where a finer distinction in the image is made is determined by the actor's input rank. However, the image of input ranks other

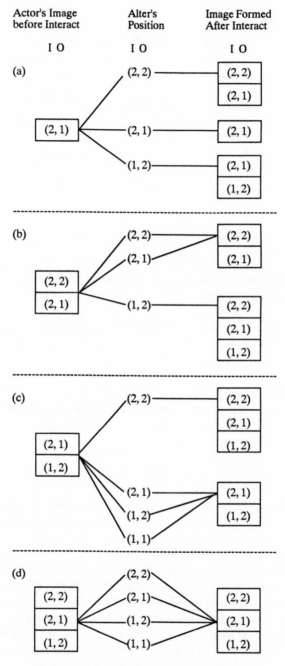

Figure 4. Development Process of Image

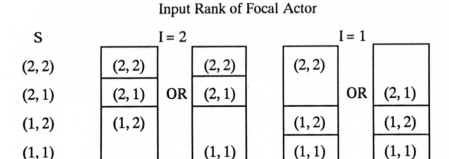

Figure 5.   Image of S in Equilibrium: Number of Ranks = 2

than the actor's own is totally dependent on the first encounter's outcome. Moreover, the image as a whole preserves the order of positions in the objective system.

The image-formation process discussed so far is based on the assumption that an actor seeks outcome information of the others at least once, however socially distant the others are (Axiom A-6). This implicitly postulates a type of society where barriers between various social groups are nonexistent, or at least low enough to allow people in different positions to examine outcome states of distant others. However, in societies where the barriers are relatively high, actors can only imagine what "those guys" get, without actually scrutinizing their outcomes.

Model Two

The section provides an alternative model for the information-seeking process and its implications for the resulting images of allocation systems. The question to be answered is the same one as in the previous section, that is, what is the the relation of the image to the actual system and the actor's position? In Model One, actors actually have access to the outcome states of the people who, in some cases, are drastically different. However, there are societies in which class barriers against interclass interactions are so strong as to be impenetrable. India's once-rigid caste system is one example. The stricter the class barrier, the harder it is for people to probe the outcome states of others beyond the barrier. In addition, human beings may be unwilling to expend the effort to seek information about dissimilar others, once they discover that "this person isn't like me."[4]

If this is the case, what happens to the image of an allocation system? We believe that actors must have some image of the allocation system even under these circumstances, simply by utilizing the information available to them. The only available outcome information is that which can be obtained without going beyond the barriers, that is, outcome information from their own kind. In short, in this section, actors are assumed to "construct" from information at hand how large a reward those dissimilar others are given.

To incorporate this idea into the framework discussed, I will introduce the following axioms to replace Axiom A-6. The first principle for constructing the outcome image is derived from findings of stratification and communication studies: We tend to direct our attention toward people of higher status and to avoid those who are of lower status. For example, Coleman and Rainwater (1978) found that the higher a respondent's social class, the higher the earnings he or she would presume necessary for membership in a given social class. Therefore, it is natural to assume that the actor will try to locate himself or herself as close to the higher status as possible in his or her image.

Axiom II-A-6-1: Actors are motivated to differentiate themselves from those of lower status, and to equalize themselves with those of higher status.

Corollary II-A-6-1: Actors tend to perceive those of higher status as close as possible to themselves. Conversely, they tend to perceive those having lower status as distant as possible from themselves.

However, it is not reasonable to assume that in forming an image, an actor totally bends to the desire to be close to the higher statuses. The objective social distance felt by the actor, although it may be distorted, should to some degree be preserved in the image. In other words, the distance between the actor and others on the input continuum should be reflected in the construction of the outcome image. Otherwise, this perceptual bias destroys the cognitive balance between input and outcome ranks in the actor's image. The basic tenet of balance theories in social psychology tells us that cognitive imbalance causes discomfort, and this motivates actors to maintain balance among cognitive elements. In line with this tenet, Headey (1991) found that the perceived and legitimate incomes of various occupations by the respondents were highly correlated ($r = .46$ to $.77$), although there was a systematic distortion in the perception of income. Since legitimate income can be interpreted as an index of input, this finding gives us direct support for the next axiom.

Axiom II-A-6-2:   Actors are motivated to maintain cognitive balance in perceptions of resource allocations.

Thus, people are motivated to perceive a balanced association of input and outcome ranks: higher input associated with higher outcome.

Corollary II-A-6-2:   Actors expect that the rank of an outcome state will correspond to the rank of the input with which it is associated.

How can the ideas described in these two axioms be combined to formalize the actor's information-seeking process? We require an axiom that will allow actors to choose the closest possible outcome as the outcome image for those above and to choose the furthest possible outcome for those below, without abusing the cognitive balance between input and outcome.

Axiom II-A-6-3:   When a focal actor X whose position is (i, k) interacts with alter Y whose input rank is j, information-seeking occurs as follows: First, X seeks information on Y's input. If Y belongs to X's rank ($i = j$), then X seeks information about Y's outcome. If Y's input is different from X's ($i \neq j$), X first locates j among the set of input ranks in his or her image. Before initiating the image-extraction process, X divides the outcome states in the image (let us call it the set O') into two subsets, based on the size of his or her own outcome: the subset of outcome states that are greater or equal to X's own outcome K ($O \geq K$), and the subset of outcome states less than K ($O < K$). Then X extracts the image of Y's outcome, O(Y), from O', the existing set of outcome ranks in the image, according to the following rules:

(1) $j > i$, where j is the p'th smallest among input ranks larger than the actor's input rank i.

O = set of the outcome states in the allocation system S
$O' \subseteq O$ = set of the outcome states in the image of actor X
set $O \geq K = \{o: o \geq k, o \in O, k \in O\}$
set $O < K = \{o: o < k, o \in O, k \in O\}$
$MinL_i$ = i'th smallest element of the intersection of sets O' and $O \geq K$
$MaxS_i$ = i'th largest element of the intersection of sets O' and $O < K$
$|A|$: number of elements in the set A
(Note that before X's image reaches equilibrium, it does not necessarily hold that $O = O'$. Any previously extracted image is corrected as X's image develops.)

a) If $|O' \cap O \geq K| = q \geq p$, then X extracts $MinL_p$ as the outcome image of Y. Thus $O(Y) = MinL_p$.

b) If $|O' \cap O \geq K| = q < p$, then X extracts $Max(O' \cap O \geq K) = MinL_q$ as the outcome image of Y. Thus $O(Y) = MinL_q$.

(2) $j < i$, where j is the p'th largest among input ranks smaller than the actor's input rank i.

a) If $O' \cap O < K = \emptyset$, then X extracts k as the outcome image of Y. Thus $O(Y) = k$.

b) If $O' \cap O > K \neq \emptyset$ and $|O' \cap O > K| = q \geq p$, then X extracts $MaxS_p$ as the outcome image of Y. Thus $O(Y) = MaxS_p$.

c) If $O' \cap O < K \neq \emptyset$ and $|O' \cap O < K| = q < p$, then X extracts $Min(O' \cap O < K) = MaxS_q$ as the outcome image of Y. Thus $O(Y) = MaxS_q$.

Figure 6 shows the general logical flow, and Figure 7 illustrates the extraction of the outcome image of input rank 2 by the actors whose input rank is 1 in a 3 x 3 allocation system.

There are two points to be noted in Axiom II-A-6-3. First, by including his or her own outcome in the subset of larger outcomes ($O \geq K$), X pulls those whose input is higher than his or hers closer to himself or herself and keeps those whose input is lower than his or hers further away. In X's image, the outcome state of the input stratum just above his or her is equated with his or her own outcome. Second, the extraction of the outcome image is based on the distances between the actor's and the alter's input ranks and between the actor's outcome rank and respective outcome states in his or her image. Consequently, the higher the input rank, the higher the outcome state assigned to it, and thus cognitive balance between input and outcome levels is maintained. This two-fold mechanism satisfies Axiom II-A-6-1 and II-A-6-2, and also obtains some empirical support (Headey 1991).

The new information-seeking process is defined as above. The rest of the model need not be changed except for Axiom A-9.

Axiom II-A-9: Given an interact, the image held by focal actor X is transformed according to the following. $P_X(Y)$ is Y's position as perceived by X. If Y's input differs from X's, then X stops the information-seeking process and extracts an outcome state from his or her current image to form the outcome image of Y according to the rule stated in Axiom II-A-6-3. The extracted outcome image of Y is denoted $O'(Y)$. If Y's input differs from X's, then $P_X(Y) = (I(Y), O'(Y))$. If Y's input is the same as X's, $P_X(Y) = (I(Y), O(Y))$.

1. if $P_X(Y)$ is already included in X's image of S, no change will occur in X's image.
2. if $P_X(Y)$ is located above the present top stratum of X's image, $P_X(Y)$ will become the new top stratum.
3. if $P_X(Y)$ is below the present bottom stratum of X's image, $P_X(Y)$ will become the new bottom stratum.
4. if $P_X(Y)$ is between two strata in the X's image, $P_X(Y)$ will be inserted between them.

Figure 8 illustrates the emergence of the image of actor X whose position is (2, 1) in a 2 x 2 allocation system S. Note that X's image of the input stratum 2 is always (1, 1) regardless of X's meeting (1, 2) or (1, 1). This is because X does not have access to the alter's actual outcome if the input rank is different, and assigns an outcome state according to the rule stated in Axiom II-A-6-3. As in Figure 4, X's image does not change once it becomes (2, 2), (2, 1), (1, 1). Being different from the previous case, no other stable image exists for the actors in (2, 1), regardless of their sequence of interaction. As a result of this process, the final images of the actors in a society depend solely on their locations in the system. Figure 9 shows the final images of 2 x 2 and 3 x 3 allocation systems.

The final images in Figure 9 share two properties with those generated in the previous section (Figure 5): that actors make finer distinctions for their own input ranks, and that their images preserve the order of positions in the actual system. However, outcome images of dissimilar others are dependent on the outcome rank of the actor, not of someone whom he or she encountered.

A substantive implication of the differences between these two image-generating processes may be interesting to consider. Let us call the former the "F-E process" and the latter the "I-E process." Images produced by the F-E process should differ for actors who have had different interaction histories, even for those in similar social positions. Although there may be typical or modal first-encounters due to an uneven population distribution among positions, there could be those who do not share the same view of the out-group's outcome with their in-group fellows. It may be difficult for them to form a solidary interest group, for example. The I-E process has a contrasting political implication, producing no variations in the image of people in similar positions.

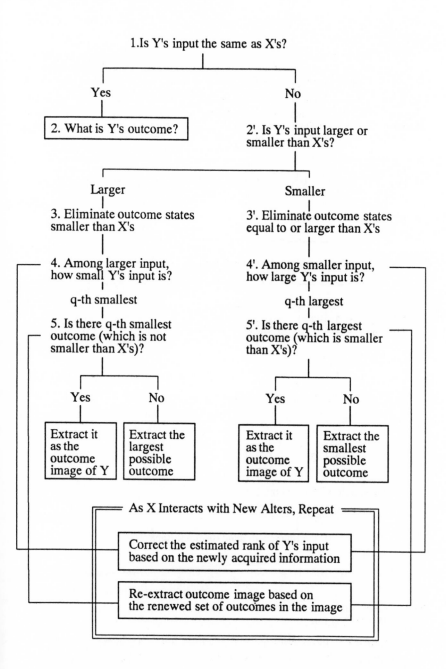

Figure 6. Outcome Image Formation Process by a Focal Actor X

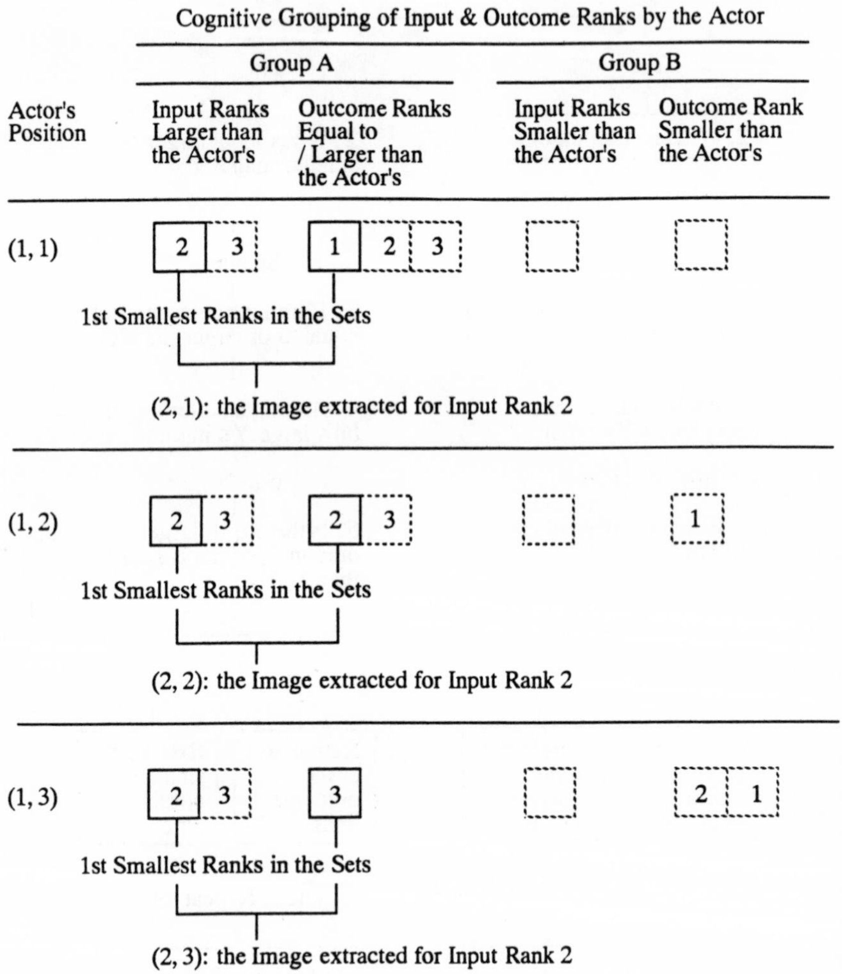

Cognitive Grouping of Input & Outcome Ranks by the Actor

	Group A		Group B	
Actor's Position	Input Ranks Larger than the Actor's	Outcome Ranks Equal to / Larger than the Actor's	Input Ranks Smaller than the Actor's	Outcome Rank Smaller than the Actor's

(1, 1) | 2 | 3 | | 1 | 2 | 3 | [ ]   [ ]

1st Smallest Ranks in the Sets

(2, 1): the Image extracted for Input Rank 2

(1, 2) | 2 | 3 | | 2 | 3 | [ ]   [1]

1st Smallest Ranks in the Sets

(2, 2): the Image extracted for Input Rank 2

(1, 3) | 2 | 3 | | 3 | [ ]   [2 | 1]

1st Smallest Ranks in the Sets

(2, 3): the Image extracted for Input Rank 2

Note: The Alter's Input Rank = 2 > 1 = The Actor's Input Rank Group B Irrelevant

Figure 7. Illustration of the Extracted Outcome Images of Input Rank 2 by the Actors of Input Rank 1 in a 3 x 3 Allocation System

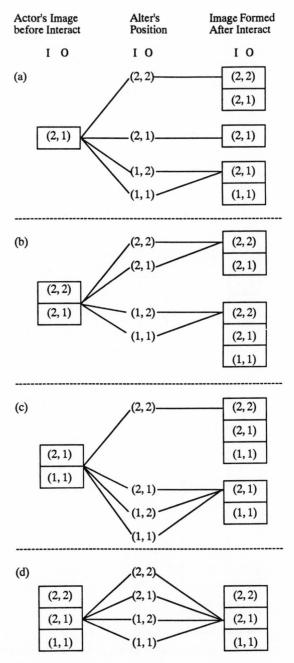

Figure 8. Development Process of Image: Model 2

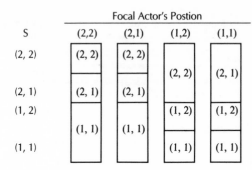

S	Focal Actor's Position (2,2)	(2,1)	(1,2)	(1,1)
(2, 2)	(2, 2)	(2, 2)		
(2, 1)	(2, 1)	(2, 1)	(2, 2)	(2, 1)
(1, 2)			(1, 2)	(1, 2)
(1, 1)	(1, 1)	(1, 1)	(1, 1)	(1, 1)

Figure 9a.   Stable Images of 2 x 2 Allocation System

Focal Actor's Postion

S	(3, 3)	(3, 2)	(3, 1)	(2, 3)	(2, 2)	(2, 1)
(3, 3)	(3, 3)	(3, 3)	(3, 3)			
(3, 2)	(3, 2)	(3, 2)	(3, 2)	(3, 3)	(3, 2)	(3, 1)
(3, 1)	(3, 1)	(3, 1)	(3, 1)			
(2, 3)				(2, 3)	(2, 3)	(2, 3)
(2, 2)	(2, 2)	(2, 1)	(2, 1)	(2, 2)	(2, 2)	(2, 2)
(2, 1)				(2, 1)	(2, 1)	(2, 1)
(1, 3)						
(1, 2)	(1, 1)	(1, 1)	(1, 1)	(1, 2)	(1, 1)	(1, 1)
(1, 1)						

Focal Actor's Position

S	(1, 3)	(1, 2)	(1, 1)
(3, 3)			
(3, 2)	(3, 3)	(3, 3)	(3, 2)
(3, 1)			
(2, 3)			
(2, 2)	(2, 3)	(2, 2)	(2, 1)
(2, 1)			
(1, 3)	(1, 3)	(1, 3)	(1, 3)
(1, 2)	(1, 2)	(1, 2)	(1, 2)
(1, 1)	(1, 1)	(1, 1)	(1, 1)

Figure 9b.   Stable Images of 3 x 3 Allocation System

FAIRNESS JUDGMENTS BASED ON ALLOCATION IMAGES

In this section I will state additional axioms that extend the image-formation model to fairness judgments. In order to make a fairness judgment about occupants of a position in the system, actors need to know two basic elements: their actual and fair outcomes. The image of the allocation system specifies what serves as the "actual" outcome in the actor's judgment, but the other element is not yet known. Moreover, to answer the third question addressed earlier in the introductory section, we need to know: (1) which positions in the image are taken into account to judge fairness of a society, and (2) how the feeling of (un)fairness should be defined. The first axiom presented in this section will deal with the conception of fair outcomes, and the rest deal with the last two questions posed above.

In the Introduction it was noted that empirical findings suggest people regard equity as a fair allocation principle. Equity demands that the size of an outcome correspond with the size of an input. In our allocation system S, this is equivalent to the positive correlation between the ranks of input and outcome. This consideration leads us to Axiom B-1.[5]

Axiom B-1 (Selection of fair outcome $O(GO_k)$): Actor X in position (i, j) chooses the fair outcome $O(GO_k)$ as follows:

1. To determine the fair outcome $O(GO_i)$ for his or her own input stratum, X examines the rank of his or her own input in the image of S, and selects the outcome whose rank equals his or her input rank. This occurs in the following way: Let $|I| = n$. If $i \geq (n + 1)/2$, then the actor counts the rank of the input from the top, and selects the outcome of equal order counted from the top. If $I < (n + 1)/2$, then the actor counts the rank from the bottom, and selects the outcome of equal order counted from the bottom. The outcome thereby selected functions as the fair outcome for stratum i, $O(GO_i)$. If there is no outcome of equal rank to the actor's input, he or she selects the input rank h that is closest to his or her own among the input ranks for which an outcome of equal order can be found. If $i \geq (n + 1)/2$, then the actor counts the rank of the input from the top, and selects the outcome of equal order with h, counted from the top. If $i < (n + 1)/2$, then the actor counts the rank from the bottom, and selects the outcome of equal order with h, counted from the bottom. The outcome thereby selected functions as the fair outcome for stratum h. This outcome, $O(GO_h)$, functions as the reference point in determining fair outcomes for the rest of the input strata, like the fair outcome for the actor $O(GO_i)$ does if there is any.

2. To determine the fair outcome for other input strata, $O(GO_k)$, $k \neq i$ (or $k \neq h$), X compares the other's input rank with his or her own input rank i, or with rank h when no outcome is assigned to his or her own input rank. X then determines how many ranks it is below or above i (or h). If input rank of the stratum k is g ranks below (above) i (or h), X selects the outcome whose rank is g ranks below (above) GO_i (or GO_h), as the fair outcome for stratum k.

The following theorem is directly derived from the above axiom.

Supplemental Theorem 1: Let $|I| = n$, $|O| = m$ and $m - n = u$. The Axiom B-1 gives as a fair reward the outcome rank which is higher than the input rank by u for the actors in the input strata equal to or higher than $(n + 1)/2$. If the input rank is i, the resulting fair outcome is $i + u$. On the other hand, actors lower than $(n + 1)/2$ have the outcome rank which is equal to the input rank as fair reward. If the input rank is i, the resulting fair outcome is i. As a special case, the fair outcome of input stratum k is outcome rank k, regardless of the actor's input rank, if the system takes the form of n x n (i.e., $m - n = u = 0$ because $m = n$).

Axiom B-1 implies that equilibration of the image of S is a necessary condition for the stability of fairness standards. Before equilibrium is attained, the numbers of input and outcome ranks in the image are continuously changing, and what is seen as the fair correspondence of input and outcome changes accordingly. Moreover, note that fair outcome is endogenously determined in the model based on the image of the allocation system. Although the resulting "fair linkage" of input and outcome satisfies the conditions specified by Berger et al.(1972)—unitary, differentiated, and balanced—this is not a referential structure in status value theory that is essentially external to actors.

With these fair outcomes in mind, how wide a range of positions in the allocation system do the actors take into account? Less motivated to probe the outcomes of dissimilar others, they may be less interested in others' fairness as they become more distant.

Axiom B-2 (Restriction of Objects of Fairness Evaluation): A focal actor X evaluates the fairness of an allocation system S by evaluating his or her own and other strata, given the following definitions:

1. Self-stratum evaluation: evaluation of whether members of X's own stratum are in a fair state and, if not, evaluation of the degree of unfairness.
2. Other-stratum evaluation: evaluation of whether the members of the strata adjacent to X's are in a fair state, and if not, evaluation of the degree of unfairness.

Underpayment has been shown to cause stronger feeling of unfairness than overpayment. This bias has been captured by taking a logarithm of the ratio of actual to fair outcomes (e.g., Jasso, 1978, 1980; Markovsky 1985; Mirowsky 1987). I will employ this method as well, partly because it is supported empirically (Jasso 1978; Mirowsky 1987). However, as will become obvious, the logarithmic terms must be summed in order to evaluate societal unfairness. Because the signs of the logarithms obtained for under and overpayment are opposite, the absolute value is employed.

Axiom B-3 (Bias in Feeling of Unfairness): A feeling of unfairness F, generated by a fairness evaluation of the situation of actor P is expressed as:

$$F(P) = |\log D|, \text{ where } D = O(P)/O(GO)$$
$$= |\log \{O(P) / O(GO)\}|$$

Axiom B-4 (Fairness Evaluation Summation): Fairness evaluations of different objects can be summed, and the evaluation of the fairness of S is equal to the sum of all the fairness evaluations made by a given focal actor.

Defining the feeling of unfairness toward the system involves a merger of the fairness judgment and image-generating processes: Actors use outcome images of each stratum as the "actual outcomes." Recall that the two different types of image formation yield different images sharing a basic property, that is, the pattern of distinctions among positions in the image. Further recall that the images of similar others—images of the self-stratum—are identical in the two processes. The point of distinction with respect to fairness judgments is the image of dissimilar others; specifically, the difference in the image of the next-higher and next-lower strata. Thus there are two formulas for F-OTHER, corresponding to the two types of information-seeking process.

Theorem 1a: A feeling of unfairness toward the n x n allocation system S is obtained by summing values of F(P) (Axiom B-4), which are obtained from the self-stratum (F-SELF) and other-stratum (F-

OTHER) evaluations (see Axiom B-2). A focal actor X in position (i, j) experiences the following degree of unfairness toward S:

F-SUM = F-SELF + F-OTHER

Theorem 1b: F-SELF and F-OTHER are defined as follows.

$$F\text{-}SELF = \sum_{k=1}^{n} |\log(k/i)|$$

$$= (2i-n)\log(i) - \sum_{k=1}^{i} \log(k) + w, \quad \text{where } w = \begin{cases} \sum_{k=1}^{n} \log(k), & \text{if } i \neq n \\ 0, & \text{if } i = n \end{cases}$$

(Note that i is the fair outcome for stratum i and there are n distinctive outcome states, ranging from 1 to n, in a stratum.)

$$F\text{-}OTHER = a \left| \log \frac{\text{outcome of stratum } i+1}{i+1} \right| + b \left| \log \frac{\text{outcome of stratum } i-1}{i-1} \right|$$

If i = n, then a = 0; and if i = 1, then b = 0. Otherwise, a = b = 1.

Theorem 1c: The Model One image-formation process gives the following:

$$F\text{-}OTHER1 = a |\log \{x/(i+1)\}| + b |\log \{y/(i-1)\}|$$
$$= a |\log x - \log(i+1)| + b|\log y - \log(i-1)|,$$

where x and y are the outcome ranks of the first encounters of the next-higher and next-lower input strata.

Theorem 1d: The Model Two image formation process gives the following:

$$\text{If } j = 1, F\text{-}OTHER2 = a |\log \{1/(i+1)\}| + b |\log \{1/(i-1)\}|$$
$$= a |0 - \log(i+1)| + b |0 - \log(i-1)|$$
$$= a \log(i+1) + b \log(i-1)$$
$$\text{If} \neq 1, F\text{-}OTHER2 = a |\log\{j/(i+1)\}| + b |\log\{(j-1)/(i-1)\}|$$
$$= a |\log(j) - \log(i+1)| + b |\log (j-1) - \log(i-1)|$$

I will next present some of the derivations from the model. First, the following propositions are common to both image-formation processes.[6]

Proposition 1: Focal actors extract the image of "fair society" from their images of S. Across actors in an n x n allocation system, the image (1) is uniform and (2) has each input stratum of a given rank receiving the equivalently ranked outcome.

In other words, there is a shared idea of "who should get what" in society if merit and reward are equally differentiated. Under this condition, everyone in the society presumes that level m merit deserves level m reward. Thus, there will be no conflict over "what is fair" in this type of society. However, this does not mean that everyone is satisfied with the society, as claimed below.

Proposition 2. Every focal actor has some feeling of unfairness regarding S.

This is because everyone sees that there are variations in the sizes of rewards received by people with the same qualification. This realization inevitably leads actors to judge society as unfair, because the same amount of reward should be given to those who are equally qualified. This comes under the issue of "inequality in conditions" suggested by Bell (1976). Although everyone finds this to be unfair, the intensity of the feeling varies with the level of qualification.

Proposition 3. F-SELF has a parabolic (high-low-high) form in terms of input, and takes the minimum value when $\log i = (n/2i) - 1$.

This implies that those having minimum feelings of unfairness will be at moderate status levels. Note that what matters in regard to the size of F-SELF is the level of qualification of the actor, not the level of outcome. It is because this model assumes that others' outcomes have the same weight as the actor's own in fairness judgments about society. This is not so strange as it may sound. Kluegel (1988) has found that personally experienced economic problems do not affect the dominant ideology—notions of how the system should be.

Recall that the Model One image-formation process does not have a stable equilibrium solution: Multiple equilibria in the image are obtained for a single position, depending on the outcomes of first encounters. This property of Model One is not necessarily desirable. We cannot obtain any one final image, that is, an actor's image of "actual outcomes" for dissimilar others. Thus we cannot know the size of F-OTHER. The following propositions are implied only by Model Two.

Proposition 4. The minimum feeling of unfairness in a given input stratum i is held by those who are fairly rewarded, that is, whose reward rank is also i.

Model Two implies that those who are fairly rewarded feel less unfair about society. However, recall that the weight given to the actor's own outcome is equal to the weight given to others' outcomes. Thus, this proposition does not imply that the actor regards society as less unfair *because he or she received* *the fair portion.* Rather, this implies that those who receive the fair reward view the world as a less unfair place. In their eyes, dissimilar others do not seem to be unfairly treated. Recall that actors extract outcome images of dissimilar others from available information, and the extraction is based on their own outcome. This process produces an image of the world that places everyone else under the same fate as the image-constructor, so to speak: if you are fairly rewarded, you view dissimilar others as fairly rewarded. Finally, Model Two predicts the pattern of the feeling of unfairness towards society as:

> **Proposition 5a.** Among the fairly rewarded, the feeling of unfairness towards society (F-SUM) takes its minimum value when the following holds:
> $$\log(i) = n/2i - (2i^2+2i-1)/[2i(i+1)]$$

> **Proposition 5b.** Among the under-rewarded, F-SUM takes its minimum value when:
> $$\log(i) = (n-i)/2i - (i^2+2i-1)/[2(i^2-1)]$$

> **Proposition 5c.** Among the over-rewarded, F-SUM takes its minimum value when:
> $$\log(i) = (n-i)/2i - (i^2-2i-1)/[2(i^2-1)]$$

CONCLUSIONS

So far I have presented the most elementary form of the model. To simplify the argument, several theoretical restrictions were set as scope conditions. In this concluding section, I wish to present a brief description of an elaboration of the theory, in order to demonstrate how the same theoretical framework can be applied to related issues in the problem of fairness and stratification.

The allocation system considered so far is restricted to be balanced in that input and outcome are equally differentiated. I assume that "balanced" allocation systems are most stable, because different numbers of input and outcome ranks connote that there are input or outcome ranks that do not have fair counterparts. An allocation system with input and outcome sets of unequal size is, by nature, somewhat unfair.

Social changes such as technological innovations or revolutions can destroy the input-outcome balance by nullifying previously meaningful distinctions or by creating new distinctions. For example, a revolution can make a formerly significant difference among ascriptive statuses nonsignificant. Analytically, we may consider the integration or differentiation of either input or outcome. Empirically, however, it seems more appropriate to speak of input becoming more integrated or differentiated.[7] Input is differentially evaluated according to the size of contribution to the society, and what constitutes a contribution depends on societal needs. In contrast, outcome does not seem to have this causal linkage with the social system.

To treat such unbalanced allocation systems, we simply delete Assumption 1. As a first step, let actors enter the system without any prior image. When input ranks are collapsed instead of expanded, the allocation system is n x m, where n < m. The following two propositions are then derived from Axiom B-1.

Proposition 6. If the input set is reduced without a corresponding reduction in the outcome set, then the standard of fair share for everyone in the society will be higher for those in the upper half of the society than in the lower half.

This proposition has an intriguing implication: that there is an agreement on "who should get what" when the merit and reward systems are equally differentiated. However, if a previously significant distinction in merit evaluation is nullified, the upper and lower halves of the society arrive at distinct ideas of "who should get what".

Proposition 7. If the input set is reduced without a correspodning reduction in the outcome set, then the idea of what the actor himself or herself deserves does not changed, but the idea of what socially distant others deserve does change.

This proposition implies that the actor's conception of fair outcome is conservatively skewed. If the actor deserved the top level outcome in a balanced allocation system, he or she still deserves the top level outcome after distinctions in the input dimension lose some of their significance.

Diffrerences in people's conceptions of "what is fair" are, of course, a potential source of ideological conflict. However, any social change that expands the input dimension instead of reducing it may have an even more serious impact on the society. Here we postulate an n x m allocation system, where n > m.

Proposition 8: When the input set expands without a corresponding expansion of the outcome set, the upper half of the society has no idea about what constitutes fair outcomes for the people in input ranks below $n - m = v$. In contrast, the lower half lacks any idea about what constitutes fair outcomes for the input ranks above $m + 1$.

Proposition 9: Let $|I| = n$, $|O| = m$, and $n - m = v$. If $v \geq m + 1$, then actors whose input ranks range from $m + 1$ to v do not know what they deserve.

It is not hard to imagine that people would feel greatly distressed when they cannot determine what they deserve. By adding axioms about the size of the distress and the selection of a strategy to reduce it, the model also implies that an allocation system in which input is more differentiated than outcome is more likely to transform itself than a system where input is less differentiated.

This discussion illustrates how scope conditions can be relaxed to deal with issues of theoretical and empirical import. Although there are many issues of sociological relevance, the process by which a specific characteristic or contribution is devalued or appreciated seems worthy of special attention. In the above, we assume that actors commence interaction in a system that is already unbalanced. However, if actors have a stable image before the transformation of the system occurs, we need to explain how they realize change in the value of specific inputs. This issue has special relevance to sociology because it concerns how objects (allocation systems) and subjects (images of the system) correspond, as well as how people deal with changes in their merit evaluations and changes in their shares—in short, how the status quo is maintained. Other directions of elaboration may include social mobility.

NOTES

1. "Linear order" is a relation that is transitive, asymmetric, and complete. Fararo (1973, pp. 73-74) provides the formal definitions of these criteria: "A relation R_A is transitive if whenever we have both xR_Ay and yR_Az, then we also have xR_Az." "The relation R_A is...complete if whenever x and y are distinct objects, either xR_Ay or yR_Ax obtains." The relation R_A is asymmetric if and only if xR_Ay implies not yR_Ax.

2. According to Fararo (1973, p. 186), an equilibrium state is defined as follows. "Let X be the state vector of a process described by a specified dynamic law. Then by an equilibrium state, denoted X_e, we mean a value of X such that under the dynamic law

if $X(t) = X_e$, then $X(t+1) = X_e$ (discrete time case)
if $X(t) = X_e$, then $X(t+\Delta t) = X_e$ (continuous time case)"

Furthermore (1973, p. 187): "An equilibrium state X_e is said to be (asymptotically) stable if, and only if, when the system state initially is not X_e, then the behavior of the system implied by the dynamic law consists in moving toward the state X_e: $X(t) \rightarrow X_e$ (as $t \rightarrow \infty$)."

3. This is because, when we assume all the interaction probabilities to be non-zero (Assumption 2), the Markov transition probability matrix for the image of any focal actor has two absorbing states. A state (in this case, an image) is said to be "absorbing" if $p_{ii} = 1$, where p_{ii} is "the conditional probability that a system in state i on the kth observation" is also in state i on the (k+1)th observation (Maki and Thompson 1973, pp. 86).

4. The reader may wonder whether or not it is possible for the actors to know the very existence of "different people" if the barriers are high. "Barriers" here are considered to be anything that hinders thorough scrutiny of the lives of those who belong to other social groups. For example, janitors working in a large corporation may have just enough latitude to know there are "executives," but they may not know how much those "executives" are earning. If they are so deprived of interaction latitude that they do not know the existence of "executives" per se, they are regarded as an isolated group. Their probability of interacting with executives is zero, and Assumption 2 excludes such cases from the theory.

5. In Axiom B-1, we presume that members of the upper half of the society locate themselves by observing how far they are from the highest contributors. In contrast, lower half members locate themselves by observing how far they are from the lowest contributors. I assume the algorithm can be justified by referring to the existence of an anchor effect (Markovsky 1988), i.e., a form of judgement bias. However, there are other algorithms for modeling the self-location process.

6. See Saito (1993) for the proofs of the propositions presented in this paper.

7. I suspect that different evaluative states of outcome are produced as a result of the association with different evaluative states of input. That is, first, input ranks are differentiated, and then outcome states acquire the ranks of inputs with which they are most frequently associated.

REFERENCES

Adams, J. S. 1965. "Inequity in Social Exchange." Pp. 267-299 in *Advances in Experimental Social Psychology* (Volume 2), edited by L. Berkowitz. New York: Academic Press.

Bell, D. 1976. *The Cultural Contradictions of Capitalism.* New York: Basic Books.

Berger, J., M. Zelditch, Jr., B. Anderson, and B. P. Cohen. 1972. "Structural Aspects of Distributive Justice: A Status-Value Formulation." Pp. 119-146 in *Sociological Theories in Progress* (Volume 2), edited by J. Berger, M. Zelditch, Jr., and B. Anderson. Boston: Houghton Mifflin.

Berger, J., M. H. Fisek, R. Z. Norman, and D. G. Wagner. 1983. "The Formation of Reward Expectations in Status Situations." Pp.127-168 in *Equity Theory: Psychological and Sociological Perspectives*, edited by D. M. Messick and K. S. Cook. New York: Praeger.

Coleman, R. P. and L. Rainwater. 1978. *Social Standing in America: New Dimensions of Class.* New York: Basic Books.

Fararo, T. J. 1973. *Mathematical Sociology: An Introduction to Fundamentals.* New York: Wiley.

Headey, B. 1991. "Distributive Justice and Occupational Incomes: Perceptions of Justice Determine Perceptions of Fact." *British Journal of Sociology* 42(4): 581-596.

Jackman, M. R. 1979. "The Subjective Meaning of Social Class Identification in the United States." *Public Opinion Quarterly* 43(4): 443-462.

Jasso, G. 1978. "On the Justice of Earnings: A New Specification of the Justice Evaluation Function." *American Journal of Sociology* 83(6): 1398-1419.
———. 1980. "A New Theory of Distributive Justice." *American Sociological Review* 45: 3-32.
Kluegel, J. R. 1988. "Economic Problems and Socioeconomic Beliefs and Attitudes." *Research in Social Stratification and Mobility* 7: 273-302.
Kluegel, J. R. and E. R. Smith. 1981. "Beliefs about Stratification." *Annual Review of Sociology* 7: 29-56.
Maki, D. P. and M. Thompson. 1973. *Mathematical Models and Applications.* Englewood Cliffs, NJ: Prentice-Hall.
Markovsky, B. 1985. "Toward a Multilevel Distributive Justice Theory." *American Sociological Review* 50: 822-839.
———. 1988. "Anchoring Justice." *Social Psychology Quarterly* 51: 213-224.
Mirowsky, J. 1987. "The Psycho-Economics of Feeling Underpaid: Distributive Justice and the Earnings of Husbands and Wives." *American Journal of Sociology* 92(6): 1404-1434.
Robinson, R. V. 1983. "Explaining Perceptions of Class and Racial Inequality in England and the United States of America." *British Journal of Sociology* 34(3): 344-366.
Robinson, R. V. and W. Bell. 1978. "Equality, Success, and Social Justice in England and the United States." *American Sociological Review* 43: 125-143.
Saito, Y. 1987. "Stratified System and Feeling of Inequity." (Kaiso-kozo to Fukohei-kan) *Sociological Theory and Methods [Riron to Hoho]* 2(1): 45-60.
———. 1993. "Modeling Individual Preception of Justice in Stratification System." Ph.D. dissertation, Indiana University. *Dissertation Abstracts International* 54(11): 4278A.
Sheplak, N. J. and D. Alwin. 1986. "Beliefs about Inequality and Perceptions of Distributive Justice." *American Sociological Review* 51: 30-46.

WHAT DOES IT MEAN SOCIAL PSYCHOLOGICALLY TO BE OF A GIVEN AGE, SEX-GENDER, SOCIAL CLASS, RACE, RELIGION, ETC.?

Theodore D. Kemper

ABSTRACT

I propose here five theoretical constructs as candidates for a model of social structure in social psychological and group process analysis. They are the *division of labor, social relations of power and status, culture, composition,* and *ecology.* Together they comprise a set that shows some promise of accounting for individual level outcomes in sociological social psychology. This is defined here as the form of analysis in which social variables are independent and psychological or individual level variables are dependent. The five constructs also promise to explain social psychologically what it means to be of a given age, sex-gender, race, and the like, what Willer and Webster (1970) termed sociological observables. I examine a sample of 100 recent articles to ascertain whether or not the sociological observables are ordinarily understood theoretically in terms of the five constructs. This proves to be so in virtually every case.

Advances in Group Processes, Volume 12, pages 81-113.
ISBN: 1-55938-872-2

Every theoretical model is a proposal for how to think about the empirical world. Its building blocks are the domain elements or concepts it takes as fundamental. Here I propose the building blocks of a model of *social structure* that I believe can deliver a productive and parsimonious representation of social processes that affect individuals, that is, a set of theoretically cogent, empirically-grounded independent variables for *sociological social psychology* in general, and for the examination of the effects of group processes on individuals, in particular.[1]

I do not closely define the term social structure here, since my purpose is not to deduce the constituents from the definition, but note only that the term structure implies a more-or-less permanent arrangement of elements that cohere and have predictable effects. The critical issue for present purposes is the selection of elements that comprise a *social* structure. As will be seen presently, the selected elements are also broadly, though unsystematically, used in present-day social psychology.

I shall proceed according to a logic proposed by Willer and Webster (1970). In a telling paper, they offered the substantial insight that sociological and social psychological knowledge will cumulate much more rapidly through the use of what they called *theoretical constructs* as opposed to more frequently used *observables*. The latter are the standard sociological categories of age, sex-gender, social class, race, religion, educational level, ethnicity, residence, region of origin, and so forth. These are the stock-in-trade, descriptive independent variables of most survey research, policy analysis, and a substantial body of social psychology.[2] However, according to Willer and Webster, observables do not lead to cumulative theory. Rather, observables must be translated into theoretical constructs. In one of their examples, the observables, occupation and sex, are both translated into the construct they call "status characteristic." Theory about status characteristics can be generalized to many more observables than occupation and sex. The conclusion is that theory would be considerably advanced were we to undertake the translation with all of the observables that we frequently use.[3]

Here I propose a set of theoretical constructs for the standard set of sociological observables.[4] I intend to explain what it can mean *social psychologically* to be of a given age, sex-gender, race, ethnicity, social class, and so forth. The distributions of these observables and the relations between them are what we most commonly intend when we refer to *the* social structure, ordinarily at the macro level, but applicable to the micro level and interpersonal relations too. If we understand the observables in terms of a productive set of theoretical constructs, we will understand also what we mean social psychologically when we say social structure and how this operates in group processes. That is, we will understand more nearly how social structures affect the individuals who are involved in them.

First I will present a vintage example of how we often resolve the problem of *not* having a minimum set of theoretical constructs signifying social structure. Second, I will propose such a set. It consists of the *division of labor, social relations of power and status, culture, composition,* and *ecology.* Applied to observables, these constructs provide a far-reaching set of denotations of the term social structure at the theoretical level. Third I will examine a sample of recent studies in social psychology for evidence of the applicability and utility of the five constructs. Finally, I will discuss some implications of using the proposed constructs and directions for future inquiry related to them.

AN EXAMPLE OF THE PROBLEM

Taking advantage of the unusually large national sample of cases (3,772 males and 2,672 females) used to standardize the well-known *16 Personality Factor Questionnaire* (16 PF) (Cattell 1973), psychologists Samuel Krug and Raymond Kulhavy (1973) published an article entitled, "Personality Differences across Regions of the United States." Given the spate of books and articles dealing with the homogenization of American culture due to the pervasive influence of the mass media, as well as the general erosion of regional differences due to the uniformity of bureaucratization, impact of national economic and social policies, and the growth of interregional travel and communication, the conventional wisdom suggested then, as now, that virtually no personality differences should be found. But, Krug and Kulhavy report:

> At the risk of oversimplifying, [the differences] might be briefly identified as creative productivity, tough-minded independence, and interpersonal isolation. The first distinguished Northeast, West Coast, and Midwest regions on the one hand from Southeast, Southwest, and Western Mountain regions on the other. The second distinguished Midwest from the West Coast and the Southwest regions. The third appears useful in describing differences between Western Mountain and Midwest regions (p. 73).

Having found regional differences in personality (28% and 30% of the variance in male and female personality, respectively, could be attributed to geography), the authors then offer two broad hypotheses (involving theoretical constructs) to *explain* their descriptive findings (involving observables). One deals with the relatively high degree of homogeneity of ethnic composition in certain regions, for example, Scandinavians in the upper Midwest, Asians and Spanish on the West Coast, and so on. The second hypothesis deals with regional differences they called, "demographic characteristics," specifically, population density, degree of industrialization, and educational opportunity. Krug and

Kulhavy suggested that the higher level of interpersonal isolation characteristic of the Western Mountain region was a function of lower population density; and that the creative productivity of the Northeast, West Coast, and Midwest arose from the greater demand for this attribute in highly industrialized areas. In general the regional patterns of personality for males and females were about the same, but where they were not, Krug and Kulhavy did not seek to explain the differences, thus failing to do for the observable sex-gender what they did for region, and thereby illustrating some of the mischief that can occur when a comprehensive understanding of the underlying theoretical properties of observables is lacking.

In the main, however, Krug and Kulhavy comported themselves sensibly: having found categorical differences in some dependent variables they looked for explanation in terms of some independent variables. In this case the dependent variables were some aspects of personality and the categorical variable was region. Krug and Kulhavy understood that the regional labels were simply observables, merely nominal identifiers, but in no sense explanatory. Thus, Northeast region is not logically related to industrialization, population density, or the ethnic and cultural provenance of its residents. Nonetheless, specific types and amounts of these potentially explanatory variables do characterize the Northeast, as they do other regions. By hunch or intuition they selected several variables—homogeneity, density, industrialization, and opportunity—that might plausibly explain what region signifies with respect to shaping personality.

To make the point general, the same explanatory vacuousness may be noted about all other descriptive or observable variables in sociological or social psychological analysis. Take race, religion, ethnicity, age, sex-gender, social class, education, occupation, and so forth. Countless studies find differences between members of different races, religions, ethnic and age groups, the sex-genders, and so on in some dependent variable or another. Thus, males are more X than females; Blacks are more Y than whites; and middle class is more Z than lower class. In each of these cases, as in the case of the regional differences in personality found by Krug and Kulhavy, it is necessary to explain the finding.

Whether the issue comes up post hoc, as with Krug and Kulhavy, or the researcher assumes at the outset that a particular observable signifies a certain theoretical construct is not material. Without judging the cogency of such attributions in particular instances, absent a comprehensive theoretical menu, the assignment of constructs to observables may be adventitious. A more systematic assignment of theoretical meaning to observables, whether post hoc or a priori, would follow from the availability of a minimum set of constructs that signify the observables. My purpose is to propose an efficient set of such constructs.[5]

Durkheim (1938) offered a useful starting point from which to generate the independent variables of sociological social psychology. In the *Rules of Sociological Method*, he defined a "social fact (as)...capable of exercising on the individual an *external constraint*" (p. 13, emphasis in original). Indeed the notions of externality and constraint have come, in general, to stand for the sociological approach to explanation. Biology and psychology are eliminated as explanation at the social level because their locus is not external to the individual. Although climate is external and does constrain, it is not a social variable. Durkheim carefully distinguished the social from the nonsocial by asserting that social facts are "every way of *acting...*" (p. 13, emphasis in original). By this we understand that social facts result from the actions of human actors—individuals, groups, or collectivities—and not meteorological forces such as climate.

I turn now to a set of theoretical constructs or variables that can comprehend the efforts of social psychologists to substruct externality and constraint and thus provide a definition of social structure.[6]

DIVISION OF LABOR

I nominate the *division of labor* as candidate for first place as a construct capable of providing social psychological explanation and reflecting social structure. The division of labor refers specifically to a more or less stable arrangement of parts, thus it satisfies the main criterion of structure set forth earlier. Of its social import there can be no doubt. As early as Plato there was serious recognition of the significance of the division of labor for social outcomes. Virtually no subsequent theory of society has omitted a consideration of the division of labor: sometimes with approval, as was true of Adam Smith, Spencer, Durkheim, Parsons, and Smelser; or with mixed feelings or disapproval as was true of Comte, Marx, Durkheim, and Weber. Some differences between these thinkers were a matter of emphasis—in particular what each specifically meant by the division of labor. Their selection of one or another aspect of the division of labor (considered as a supervariable, see Note 6) conditioned their approbation or disapprobation. However, what was common to them all was the crucial point that variability in the division of labor in some sense is critical for social psychological outcomes, and therefore, in an important sense, *explains* them.[7]

For the purposes of sociological social psychology, or for sociology, broadly speaking, the division of labor has a compelling attraction as an analytical point of departure. Indeed, humans comprise a social species precisely because they cannot satisfy or accomplish alone a broad panoply of tasks and interests. Thus,

we are necessarily social, and to be social involves cooperation in some form
or another in a division of labor, or as Marx (1964, p. 62) put it, "by social
is meant the cooperation of several individuals, no matter under what
conditions, in what manner or to what end." Though Durkheim (1933) devoted
one of his great works to the division of labor, he seems to have missed this
fine point about cooperation in his discussion of mechanical solidarity. Even
there, cooperation, which involves a division of labor, was a necessity for social
survival, and Weber (1947) proposed a form of the division of labor—the
parallel type—that fully accounts for the cooperative processes in Durkheim's
mechanically solidary societies (see also Kemper 1975).

In its multiple and various aspects the division of labor implies an
organization of social effort and, with regard to any individual involved in it,
both the *whole pattern* of the division of labor and the individual's *position*
in it are exterior and constraining. It thus qualifies as a social fact in the
Durkheimian sense and as an independent variable for social psychological
explanation. If observable social categories differ on some dependent variable,
social psychologists may look first to the division of labor for explanation.
This does not mean that the division of labor will always explain the obtained
differences. Rather, we may begin with the division of labor en route to
explanation.

As a supervariable, the division of labor stands for a whole family of
subvariables and indicators. Each, however, can be understood as a form or
aspect of the organization of technical activity that actors participate in together
in order to attain ends and goals that they cannot at all or as feasibly attain
alone.

I propose three broad forms of analysis that can be applied when the division
of labor is used as an explanatory variable. First is a view of the division of
labor from an economic perspective; second, from a network perspective; and
third from a job or task perspective.[8]

Economic Properties of the Division of Labor

Although apparently little known, Weber (1947) provided a useful set of
formal properties of the division of labor when it is viewed as a system of
production. He set forth the following.

1. The manner in which an individual is involved in the division of labor.
The individual may (a) perform different technical activities all leading to a
single product, as a craftsperson does. This Weber called *specification* of
function. Or (b) the person may perform only a single technical function that
must be combined with the work of others to create a product, for example,

in an assembly line, or in a university. This, Weber called *specialization* of function. Clearly, we have in the distinction between specification and specialization a designation of the differences in labor that Marx (1964, pp. 167-77) made much of as a source of alienation in the transformation to capitalism.[9] A more recent treatment of the same question was Blauner's (1964) examination of Seeman's (1959) dimensions of alienation in four industries representing different degrees of specification and specialization of labor. The work of Kohn (1977) and his colleagues is also an important examination of the same issue. We can see too that specialization at the organization level is one of the main features of bureaucratization (Weber 1946), and, at the societal level, it is one of the main conditions of industrialization (Smelser 1959). Weber continues:

2. The entire division of labor may take different forms. (a) The common task may be accomplished by an "accumulation of functions." This may entail either (1) a *parallel* form, in which all participants are doing the same thing, for example, a crew of apple pickers who together harvest the crop, or the residents of Durkheim's mechanically solidary societies. Or (2) they may be linked in a *single common purpose* to achieve the goal, for example, all pulling on a rope to drag an extremely weighty object like the blocks of stone that Egyptian slaves hauled to build the pyramids.
   Instead of accumulation, the common task may be accomplished (b) through a "combination of functions." This includes (1) the *successive* type, for example, an assembly line and (2) the *simultaneous* type of combination of functions as in the performance of a symphony orchestra, or making love, or waging war. Weber also proposed a *managerial* function that is either separate or in addition to involvement in other functions.

Although Weber examined the division of labor under the rubric of "sociological categories of economic action," it is clear that as forms of task organization they apply in all institutional domains, whether at the macro or micro level, and consequently can have social psychological effects on individuals who are involved in different types of positions and whole patterns of cooperative, that is, divided, labor.

Network Properties of the Division of Labor

A second approach to the division of labor is through its network properties. This approach was stimulated by early work with communication networks (Bavelas 1948; Leavitt 1951; Berkowitz 1956). Plausibly, communication nets and the division of labor are coterminous. Cooperation necessitates communication about the task, its processes, allocation of responsibilities, and so forth. Hence, how communication is effected—literally, who talks to whom about the task—is a template of the division of labor. Updated, the conception in much current work with networks (e.g., Markovsky, Skvoretz, Willer, Lovaglia, and Erger 1993) is who exchanges with whom. Exchange is a general construct that embraces any number of types of cooperative endeavor in a division of labor.

With this understanding, formal, or graph-theoretic, properties of networks become of interest, since they represent also formal properties of the division of labor as a whole and the individual positions in it. For example, is there a central position? Are there channels of communication between all occupants of positions? What is the average degree of peripherality. With respect to the actor's location in the division of labor, is it a central or a peripheral position? How many connections does it have with other positions? Is the position at the beginning, in the middle, or the end point of a process? Is it a cutpoint, which means that the system will fail if the function in that position is not performed (Kemper 1972). Indicative of the growing edge of the division of labor construct is the recent notion of "structural hole" (Burt 1992) which signifies an omitted link in a network. Burt argues that such holes leave some actors significantly advantaged in their interactions with other network members.

One of the most compelling social psychological studies involving the division of labor conceived as a communication network was a carefully done early experiment by Berkowitz (1956), who pitted individuals' personality disposition against their division-of-labor position to determine which would prevail in determining behavior over a series of trials. In a quite striking result, the division-of-labor position predominated over personality by the third trial of a cooperative task. In recent network studies, a significant interest is in how different network arrangements promote power advantages to different positions (e.g., Markovsky, Willer, and Patton 1988), an issue to which additional attention will be directed subsequently.

Job Properties of the Division of Labor

A third type of division of labor analysis is concerned with the properties of the position conceived as a job or task. Does the position deal with symbols, people, or things (Kohn 1977). Does it involve supervising or being supervised? Is the end-product considered socially useful? Kohn (1983) and his colleagues (Kohn and Schooler 1982; Schooler 1987) have developed a set of analytical properties of positions in the division of labor with particular attention to the organization of the technical activity that is required by a particular position, for example, substantive complexity, routinization, and opportunity for self-direction (Parcel and Menaghan 1990). Social psychological studies using this type of division of labor analysis have linked job conditions to intellectual flexibility (Kohn 1983), psychological distress (Parcel and Menaghan 1990), and self-esteem (Gecas and Seff 1990) among other dependent variables.

These three forms of the division of labor do not exhaust the construct, but rather provide an overview of the several directions the supervariable has taken in recent years. Overall, it is no accident that the great classical theorists of sociology valued the division of labor as an analytical tool. They understood too that some of the most important processes in society were linked to the division of labor, for example, stratification and socialization, and that both group and individual fate were significantly linked to location in the division of labor. For present purposes, the division of labor serves as the primary theoretical construct by which to understand what it means social psychologically to be of a given race, sex-gender, religion, and so forth. If we wish to explain the differences in behavior, attitude, emotion, motivation, choice, selection, status, power, attainment, and so on of members of these categories, we may look at the patterns of division of labor as a whole and the kinds of positions the members of the different subcategories inhabit.

As will be seen, social psychologists and group process analysts already do this, but somewhat unsystematically. With a core set of division of labor constructs to work with, investigators can more easily theorize beforehand what kinds of division of labor effects may be operating on their dependent variables.

SOCIAL RELATIONS OF POWER AND STATUS

The second sociological supervariable I propose is linked in many theories to the division of labor as one of its most significant outcomes. Whether this linkage is accepted or not, it is possible to view the supervariable of *social relations* in the following way: actors playing their parts in the division of labor are engaged in technical activities that summate in some fashion, as discussed earlier, to attain a goal that is impossible or infeasible to attain by the actors in isolation from each other. Human actors, however, orient their activities to each other for more than technical ends. Actors relate to each other with approval and disapproval, love and hate, domination and defiance, and so forth. They are thus bound to each other socially not only by their interdependence in the division of labor, as described earlier, but also in terms of social relational activity that is responsible for both positive and negative affective outcomes between actors. Importantly, social relations between actors depend partly on the nature of their involvement in the division of labor. Thus central and peripheral actors in division of labor networks are likely to experience different social relations with other actors.

A significant body of research displays an unusually high degree of convergence on two central dimensions of social relational—as opposed to the division-of-labor—activity of human actors (Kemper and Collins 1990). Variously named, the dimensions appear to suggest two basic constellations: one is a set of relational behaviors that entail threat, control, punishment, or rejection, and are associated with actual or attempted domination in social relationships. Weber (1946) captured the essence of this pattern, through which one actor attempts (and often succeeds) in obtaining his or her own will over the resistance of other actors, in the term *power*. When power is used to gain compliance the compliance is involuntary.

The second important relational pattern is a set of behaviors that entail support, approval, friendship, sociability, gratification, inclusion, acceptance, reward, and, at its ultimate, love. Expressed in manifold ways, compliance with the wishes, desires, needs, and requests of another actor is given voluntarily. Kemper and Collins (1990) referred to this constellation of behaviors as *status-accord*, or status in brief.

It is not difficult to conceive of the power and status dimensions as constituting axes of relationship along which the members of different races, religions, sex-genders, educational levels, occupations, and so forth are positioned vis-à-vis each other. Since patterns of power and status relations tend toward relative stability, they constitute a social structure, in which the elements are members of social categories and the lines of connection between them the social relations that indicate the degree of voluntary and involuntary compliance that links them.[10]

One of the most pervasive instances of power and status effects is in the domain of social class analysis. If we find behavioral, attitudinal, motivational, and other differences between middle-class and working-class individuals, we can look, of course, to the location of these classes in the division of labor, as Marx (1964) did early and Kohn (1977) has done more recently. We can also examine class differences in light of Weber's (1947) economic analysis of the division of labor, as detailed previously. But we may also look, as Weber (1946) proposed, at the different power and status positions of the classes in question. Indeed, power and status are what interaction between social classes is about, in the relational sense (cf. Kemper and Collins 1990).

Several important traditions in contemporary social psychology and group process analysis are based on the power and status dimensions. Expectation states or status characteristics theory is concerned with the power and status implications of social category membership (Berger, Fisek, Norman, and Zelditch 1977; Berger, Fisek, Norman, and Wagner 1985; Ridgeway and Berger

1988). To be of a given age, sex-gender, race, and so forth—called *diffuse* status characteristics in this work—is to evoke differential opportunity to make task contributions, differential judgments of task competence, and differential influence over group decisions. A second tradition of research, especially in the power domain, stems from the important work of Emerson (1962) on power-dependence relations and, in one branch, links also to equity theory (Cook and Emerson 1978; Cook, Hegtvedt, and Yamagishi 1987); in another branch, research has focused on power as a feature of network position producing advantages or disadvantages in exchange relations (Molm 1985, 1987; Markovsky, Willer, and Patton 1988; Willer 1987; Markovsky, Skvoretz, Willer, Loveaglia, and Erger 1993).

A third social psychological tradition in the power/status domain examines these relational elements as determinants of emotions. Heise (1979) and Smith-Lovin and Heise (1988) have looked at emotions from the perspective of the semantic differential dimensions of potency, evaluation, and activity. These are arguably analogues of power, status, and the division of labor, respectively (see Kemper and Collins 1990). Collins (1975) has focused on the emotions of order-givers and order-takers, that is those with higher and lower levels of power and status in organizations. Kemper (1978, 1991) examined outcomes of power and status relations as determinants of a variety of human emotions.

Power and status constitute a structure of relations between individuals as members of different formal or informal social categories. Intergroup relations can be said to consist largely of such patterns of voluntary and involuntary compliance. Using a regression-analysis metaphor, after variance is independently accounted for by the division of labor, the power and status distribution may explain further what it means social psychologically to be of a given social category.

CULTURE

The third type of social variable that entails exteriority and constraint for social psychological purposes is *culture*. For Durkheim, culture was the social fact par excellence. Collective representations or conscience (Durkheim 1933) is essentially what we understand today as shared, learned, and transmitted (i.e., cultural) values, norms, and standards. Two aspects of culture in respect to social structure are important: content and consensus.

Culture Content

Culture content directly contributes to the formation of social structure through its specification of several important structural elements:

1. Through technological knowledge, culture makes possible the division of labor as a whole, as well as the individual positions in it. Marx (1964, p. 93) put this most trenchantly when he wrote: "The organization and division of labor varies according to the instruments of labor available. The hand mill implies a different division of labor from that of the steam mill." Culture also provides the labels that identify the different positions to which tasks are assigned.

2. Culture prescribes the power and status distance between positions in the division of labor—physicians have more of both than clerks, teachers more than farmers, fathers more than mothers. This may be after the fact, so to speak. That is, a particular position may have gained its status standing after a successful assertion of power which, when enforced long enough, may lead to cultural legitimation.

3. Culture defines how much power and status particular behaviors signify, for example, the gradations of verbal excess that may be construed as insult or the degree of attention that may be understood as authentic status conferral. In American society, a gift of red roses signifies affection. In Eastern satrapies of old, the prone position signified what was due in status to the ruler from his subjects. Regardless of size, each distinct social group develops a somewhat unique codebook of meanings of acts in terms of their relational implications. Culture also specifies what kinds of behavior are worthy of status-accord. For example, high education and occupational achievement in advanced Western societies and trance-like states in some primitive societies.

4. By means of socialization, culture ascribes to the different social categories (the observables of age, sex-gender, race, etc.) different division of labor positions and their respective power and status standing—for example, males > females, whites > nonwhites, adults > children. All social distance scales (e.g., Bogardus) reflect the cultural opinion of different social groups about the status standing of other groups relative to themselves.

5. Culture provides values and belief systems that shape rationales and justifications—ideologies—either for maintaining or for changing the social structure. That is, for retaining or revising the existing division of labor, the power and status value of its positions, and the assignment of social categories to their respective power and status levels.

In sum, different races, religions, sex-genders, classes, and so forth differ in part because they are assigned different division of labor positions and their associated power and status levels. Cultural content that is shared, learned, and transmitted to the various social categories shapes their behavior, their beliefs, and their degree of acquiescence in the existing social structure. The Functionalist school in sociology, exemplified by Parsons (1951, 1967), has favored culture-level explanations of stable social structures. Krug and Kulhavy resorted to this level when they suggested that personality differences between regions might be due to heavier concentrations of certain ethnic groups—with their distinctive cultures—in certain regions of the United States.

Consensus

A second analytic component of culture that bears on social structure is the degree of consensus on the various elements of content discussed earlier. This is one of the debating points between Functionalist and Conflict theorists in sociology (Mann 1970; Turner and Abercrombie 1980). In social psychology, a consensus notion has been less likely to stir turbulence. Yet it has some currency. Mead's (1934) "generalized other" is a consensus concept, as is Durkheim's (1933) "collective conscience." Blumer (1937) proposed that social dissensus leads to individual disorganization, although it can also lead to innovation and creativity. Stanton and Schwartz (1954) found that staff dissensus affected patient well-being in mental hospitals. In the classic experiment by Asch (1956), high consensus in the social milieu on an erroneous cognitive judgment constrained a substantial minority of persons to make, or exhibit external conformity to, the same error. Recent studies have examined social consensus as a feature of group formation (Insko, Pinkley, Haring, and Holton 1987), emotions (Rosenberg 1990), and consumer confidence (White, Tashchian, and Ohanian 1991).

If between-group differences can be explained in part by differences in culture content, within-group differences are partly a function of consensus on the content promulgated by dominant social categories for the consumption of all groups. Although lack of consensus may seem to imply an alternative content, for example, of values, norms, and standards, these may not be articulated with the same degree of concreteness and elaboration as the elements of the dominant culture.

In sum, differential participation in social structures reflecting the division of labor, power and status relations, and culture can be *directly* constitutive of the social psychological differences between members of different races, religions, age groups, the sex-genders, and so forth. Two additional variables, to be discussed now, are in some sense either preconditions or consequences of certain patterns of the division of labor, power and status relations, and culture.

As pointed out by Willer and Webster (1970), the standard set of sociological observables consists of categories of actors aggregated under a common rubric, for exammple, age, sex-gender, race, and the like. There are in addition two observables that refer to aggregates of actors not in terms of their categorical memberships, but in terms of the *compositional* and distributional or *ecological* features of the groups under examination.[11] Although compositional and ecological features of groups are also exterior and constraining, I believe they obtain their effects indirectly through one or more of the construct level variables previously discussed.

COMPOSITION

Composition addresses the question of who, in the sense of social position, constitutes the social environment. One of the principal compositional variables is group size. This is one of the most frequently examined group variables, perhaps because it is one of the easiest to measure. Notwithstanding ease of measurement, it is important. All social psychological analyses of rural vs. urban effects imply that size makes a difference. Easterlin (1980) has spun a wide-ranging theory concerning the fate of members of birth cohorts that differ in size. The well-known "marriage-squeeze" (Guttentag and Secord 1986) resulted from unequal numbers of males and females born into successive cohorts during the population of the baby boom years. Given the cultural propensity of older males to marry younger females, the later cohorts of females outnumbered the earlier cohorts of males. The so-called "baby bust" of the period beginning in the mid-1960s reversed the consequences with fewer females in the later cohorts available for the larger number of older males. In recent work, group size has been shown to affect helping behavior (Amato 1990), loneliness (Kraus, Davis, Bazzini, Church, and Kirchman 1993), and motivation (Kameda, Stasson, Davis, Parks, and Zimmerman 1992).

Presence or absence of a particular social type, that is, someone who occupies a specific division of labor position or engenders certain power and status relations or provides certain cultural prescriptions, is another important compositional variable. At the macro level, societies lacking a stable middle class of sufficient size are said to be poor candidates for stable, democratic regimes (Lipset 1960). At the micro level, sex of siblings affects sex-gender-role acquisition (Brim 1958). Garfinkel and McLanahan (1986) found that women growing up in a home without both biological parents were more likely to have an out-of-wedlock child. Living in a neighborhood that is religiously consonant or dissonant with one's own religious identity affects self-esteem and psychological distress (Rosenberg 1962), while neighborhood racial consonance affects Blacks' identification with other Blacks (Lau 1989).

Who the others are in the group affects the availability of prescriptions or models for cultural content; or provides greater or lesser power-and-status relational experience with these types of socially defined others; or entails a different type of participation in the division of labor. Thus, differential composition of the social context as this varies systematically by subgroups such as race, religion, age groups, the sexes, and so forth, to a certain extent determines each group's fate. What is called "contextual analysis" or "structural effects" also entails the compositional variable.

ECOLOGY

Ecology refers to the distribution of actors in physical space relative to any focal person(s). The work of Hall (1959) on cultural differences in appropriate physical distance during social interaction as this varies in different societies is pertinent here, as is the work by Festinger, Schachter, and Back (1950) on the effects of location in a housing project on sociability patterns of families. Strodtbeck and Hook's (1961) analysis of position around the jury table as this affects positive regard and sociometric choice behavior between jurors also pertains to the ecological variable, as does the work of Sommers (1959) on the relation between seating and interaction. Many investigators, including Krug and Kulhavy (1973), have been concerned with population density, which is an interaction between the compositional variable of size and the ecological variable of distribution in space. At the macro level, the interaction effects of size, division of labor, and centralization of power have been noted by many (Durkheim 1933; Hawley 1950; Mayhew 1973). At the micro level, density or crowding has been examined frequently in relation to psychological distress (Lepore, Evans, and Pulsane 1991; Lepore, Evans, and Schneider 1992; Ruback and Pandey 1992).

This concludes the presentation of the model of social structure that I propose embraces efficiently the broad spectrum of independent variables in sociological social psychology and group process analysis.[12]

EXAMINING THE MODEL

In order to examine the comprehensiveness of the proposed model[13] of social structure, I have constituted a sample of 100 recent articles, moving backward from the March 1993 issue, published in the *Social Psychology Quarterly*, which is the premier social psychology publication for sociologists.[14] Although this method of sampling is arbitrary, I am not seeking to estimate parameters, but rather to obtain a sense of goodness of fit between the set of proposed constructs and what investigators are already doing in current social psychological research.[15]

Table 1. Sociological Observables as Theoretical Constructs

Observable	Theoretical Construct	Operationalization
AGE		
Corsaro (1992)*	Power	"adult power and authority" over children (p. 171)
EDUCATION		
Schuman, Bobo, and Krysan (1992)	Power	Agreement with F-scale items (in middle class only)
Parcel and Menaghan (1990)	Division of Labor	Occupational complexity (19 items)
Stevens, Owens, and Schaefer (1990)	Power and Status/Culture	Socioeconomic status/Values & attitudes
ETHNICITY		
Learner and Grant (1990)	Status	"minority children have lower status" (p. 230)
Riches and Foddy (1989)	Division of Labor/Status	Performance evaluation/Initiate action & exercise influence
MARITAL POSITION		
Scanzoni and Godwin (1990)	Power/Status	Control/Love & income
NATIONALITY		
Feather and McKee (1993)	Culture	Self vs. collectivity values pertaining to achievement
Schwartz, Struch, and Bilsky (1990)	Culture	Rokeach (1973) Values
RACE		
Simon, Eder, and Evans (1992)	Division of Labor	Female involvement in marriage (whites) vs. work (blacks)
Young (1991)	Power	Quality of experience with police
Demo and Hughes (1990)	Culture/Power & Status/ Composition	Parental values/Socioeconomic status/Identity of interaction partners
Hallinan and Kubitschek (1990)	Status	"higher Status is assigned to white than blacks (p. 253)

(*continued*)

Table 1.   (Continued)

Observable	Theoretical Construct	Operationalization
RACE (Continued) Felmleee, Spreacher, and Bassin (1990)	Culture	"same race or ethnic back- ground have more in common" (p. 28)
Sev'er (1989)	Status	Influence
RELIGION (No studies in this category).		
RESIDENCE (No studies in this category, e.g., rural vs. urban)		
SEX-GENDER Fox and Ferri (1992)	Culture/Status/Power	"sex [affects] expectations, pres- tige, power," (p. 267)
Kane (1992)	Power	Males are socially dominant (p. 311)
Moore (1991)	Status	"being a woman means having lower...status (p. 220)
Sprecher (1992)	Power/Culture/Divison of Labor	"men...have more powerful positions (p. 58)/sex role sociali- zation (p. 58)/women more relationship-oriented (p. 59)
Shanahan, Finch, Mortimer, and Ryu (1991)	Culture	"differential...moral orientations and values" (p. 303)
Holtgraves (1991)	Power & status	Females interpreted replies to questions "indirectly" (p. 22)
Hegtdvedt (1990)	Status	"males typically occupy the higher status position" (p. 218)
Hallinan and Kubitschek (1990)	Status	"higher status is assigned to males than to females" (p. 253)
Jackson and Sullivan (1990)	Division of Labor	"work roles [imply] masculinity, family roles [imply] femininity (p. 279)

(continued)

Table 1. (Continued)

Observable	Theoretical Construct	Operationalization
SEX-GENDER (Continued)		
Naoi and Schooler (1990)	Division of Labor/Culture	Women's jobs less self-directed/ Norms keep women subservient (p. 100)
Scott (1989)	Culture	Women tied more closely to groups espousing traditional norms (p. 324-25)
SOCIAL CLASS		
Whitbeck, Simons, Conger, Lorenz, Huck and Elder (1991)	Division of Labor/ Power & Status	Work history/Income and family debt
Wiltfang and Scarbecz (1990)	Division of Labor/ Composition	Employment/Neighborhood employment rate

Note: * Citations are found in Appendix

Table 1 shows how ten common sociological observables, from age to social class, were treated at the construct level by investigators whose work was included in the sample.[16] In addition, to support my judgment that the observable is treated in terms of one of the proposed constructs, I have provided a measure or a summary of a discussion or a direct quotation, linking the observable and the investigator's understanding of it in construct terms. For example, Corsaro (1992) proposed that age differences (between parents and children) can be understood as power differences and this is expressed as a theory statement. In another example, Whitbeck, Simons, Conger, Lorenz, Huck, and Elder (1991) construed social class in terms of the division of labor and power and status. These were operationalized by measures of work history, income, and family debt.

The distribution of studies of observables is quite uneven. Some of this is probably due to the fact that specialized journals for the given observable preempt the publication of many articles dealing with them in *Social Psychology Quarterly*. For example, only one article (Scanzoni and Godwin 1990) deals with marital position; several specialized journals are devoted to marriage and family. On the other hand, sex-gender, currently a very popular topic across the sociological spectrum, has more studies pertaining to it than any other observable. Neither religion nor residence (understood as urban vs. rural) can be found as descriptive independent variables in any of the studies in the sample. These variables, once so powerful in sociological analysis, signify relatively little in contemporary American society, at least as far as authors in the sample see it. Religion has long ceased to be a distinctively divisive cultural category (Warner 1993). And with the growing urbanization of American society, the search for residential differences is more futile than meaningless.[17]

Yet, we must be sensitive to the issue of fashion in social psychological analysis. A polar opposite of the declining real significance of religion and residence is the increasing importance of sex-gender. When religion and residence were important observables sex-gender was important too, but it was not fashionable to examine this variable. Today, sex-gender remains as important as ever but fashion has changed to include it among the most studied observables.

In every case investigators construed their observable(s) in terms of one or another of the constructs of social structure proposed here, except when they omit or abandon social analysis entirely. For example, Pallas, Entwisle, Alexander, and Weinstein (1990, p. 313) speculated that sex-gender differences in adolescents' judgment of their appearance might be due to girls experiencing bodily changes earlier than do boys. In some cases more than one of the theoretical constructs was appointed to represent the observable, for example, Riches and Foddy (1989) selected both division of labor and status to stand for education.

Different investigators, doubtless with different interests in mind, selected different theoretical elements to represent the observable they were dealing with. In regard to race, the division of labor, power, status, culture, and composition appeared in the six studies in the table.

Although many investigators continue to work at the level of observables, it can be seen from Table 1 that virtually all must ultimately resort, as did Krug and Kulhavy (1973), to theoretical constructs to generate explanation of their findings. However, some investigators work with theoretical constructs *ab initio*. Table 2 shows these studies from the sample drawn from *Social Psychology Quarterly*.

Studies are listed under the heading of their appropriate construct in the social structure model. The operationalization of the construct and the dependent variable to which its use is directed are also shown. It is not astonishing to find that many studies reference the division of labor. Power and status are also presently popular constructs. Somewhat unexpected was the strong showing of composition. Yet, since this is a surrogate variable for the division of labor, power and status, and culture, this might have been anticipated. No similar anticipation would apply to ecology, a once popular variable that seems to have lost its cachet and has not a single study in the sample of the recent articles from *Social Psychology Quarterly* reflect it.[18]

Table 2. Theoretical Constructs in Use in Sample Articles

Construct	Operationalization	Dependent Variable(s)
DIVISION OF LABOR		
Ross and Mirowsky (1992)*	Employment	Sense of control
Thoits (1991, 1992)	Mix of social roles	Psychological well-being
Michaelson and Contractor (1992)	Position in social network	Perceived interpersonal similarity
Smith (1990, 1992)	Teaching younger siblings	Growth in academic achievement
Shanahan, Finch, Mortimer, and Ryu (1991)	Occupational self-direction & skill demands	Depressive affect
Skvoretz and Willer (1991)	Centrality of position	Favorability of outcomes in exchange
Friedkin (1990)	Network position	Interpersonal influence
De Gilder and Wilke (1990)	Task performance	Influence
Smith (1990)	Teaching younger siblings	Reading and language achievement
Moen and Forest (1990)	Autonomy with regard to absences	Psychological strain, mental & physical exhaustion
Parcel and Menaghan (1990)	Mother's hours, pay & substantive complexity of work	Child's cognitive abilities
Gecas and Seff (1990)	Occupational complexity	Perception of self competence & worth
Mortimer and Lorence (1989)	Work autonomy	Job satisfaction
POWER		
Ford and Blegen (1992)	Punitive tactics in bargaining	Retaliation with a punitive response
Mazur, Booth and Dabbs (1992)	Winning a chess match	Testosterone level

(continued)

Table 2.   (Continued)

Construct	Operationalization	Dependent Variable(s)
POWER (continued)		
Hegtvedt (1990)	Control over benefits to self or other	Emotions
Kemper (1991)	Domination, control or punishment by self or other	Emotion
Komorita, Aquino, and Ellis (1989)	Bargaining strength	Gain from investment
STATUS		
Stewart and Moore (1992)	Pay	Resistance to influence
Kemper (1991)	Voluntary conferral of benefits by self or other	Emotions
Hegtvedt (1990)	Higher or lower social evaluation	Emotions
Scanzoni and Godwin (1990)	Love for other	Satisfaction with negotiation outcome
CULTURE		
Stasson, Kameda, Parks, Zimmerman, and Davis (1991)	Type of consensus on rules for group decision-making	Learning and performance
Felmlee, Sprecher, and Bassin (1990)	Consensus of opinion in social network	Dissolution of premarital relationship
Scanzoni and Godwin (1990)	Consensus on negotiation outcome	Satisfaction with negotiation outcome
Feather and McKee (1993)	Values	Attitudes toward high achievers
Schwartz, Struch, and Bilskey (1993)	Value similarity/Value stereotypes	Allocating resources to ingroup vs outgroup
COMPOSITION		
Jackson (1992)	Spouse & friend in support network	Psychological strain
Insko, Schopler, Kennedy, Dahl, Graetz, and Drigotas (1992)	Working alone or with others	Cooperation or selfishness

(*continued*)

Table 2.   (Continued)

Construct	Operationalization	Dependent Variable(s)
COMPOSITION (continued)		
Krauz, Davis, Bazzini, Church, and Kirchman (1993)	Size of friends network	Loneliness
Stasson, Kameda, Parts, Zimmerman, and Davis (1991)	Problem-solving experience in group	Problem-solving quality alone
Grube and Morgan (1990)	Substance use by friends	Substance use by self
Rabow, Newcomb, Monto, and Hernandez (1990)	Presence of friends or family	Intervention in drunk-driving situation
Wiltfang and Scarbecz (1990)	Social class of neighborhood	Adolescent self-esteem
Lau (1989)	Proportion of group members in local environment	Identification with group
Amato (1990)	Size of social network	Helping behavior
Kamada, Stasson, Davis, Parks, and Zimmerman (1992)	Subgroup size	Motivation
ECOLOGY (No studies in this category)		

Note: * Citations are found in Appendix

DISCUSSION

My purpose here was to provide a set of theoretical constructs to populate a model of social structure for sociological social psychology and group process analysis. Each of the constructs—division of labor, social relations of power and status, culture, composition, and ecology—has been employed recently or in the past to explain differences in observable variables such as age, sex-gender, race, and the like. Social structure indicates some stable arrangement of elements that have social relevance and the five constructs comprise the tissues and sinew that bind observables into a structure.

Now that five structural elements have been presented, it is useful to recapitulate some earlier views. For example, Blau (1975) defined social structure as:

the differentiated social positions or statuses in a collectivity and the role relations of people as influenced by their statuses.... Social structure is delineated by numerous parameters that differentiate social status along various lines (p. 131).

Clearly, the model proposed here sets out the differentiation of social positions, via the division of labor, and delineates the various parameters of differentiation by the remaining four constructs in the social structure model posited, thus giving substance to the otherwise uninformative "various lines." Nadel (1964) reviewed earlier efforts to define social structure and it is instructive that with only a single exception, all provide only empty, formal statements about such matters as arrangements of parts. The exception is Leach (1954), who conceived of structure as a "distribution of power between persons or groups of persons" (p. 4). This, at least, gets at some of the content that is arranged in any *social* structure.

To provide a set of elements of a social structure model in no way settles any final issues. Some of the questions that remain very much on the agenda are the following.

Since each element in the model is a supervariable, there are many subvariables that pertain to it. New subvariables will continue to be discovered, for example, Burt's (1992) structural holes in the network approach to the division of labor. This does not disturb the model, but rather will help to enrich it. When investigators detect a new subvariable, they can more readily conjecture concerning its validity because a supervariable framework exists within which to examine the new concept. Indeed, one of the most important contributions of the social structure model proposed here is that it will enable investigators to articulate new theoretic entities and measures within an existing framework. The framework provides a basis for an examination of convergent and divergent validity (Campbell and Fiske 1959).

Nor is this to say that entirely new constructs at the level of the division of labor will not be discovered. That possibility does not affect the current validity of the proposed model of social structure.

The social structure model does not end inquiry, but rather initiates it at a new level, bringing us closer to true social scientific explanation. For example, if occupational complexity influences intellectual flexibility (Kohn 1983), we must go on to ask how it does this, what mechanisms operate, and so on, thus gaining insight into the kind of integration of relevant social experience and individual psychology that sociological social psychology promises to supply. Schooler (1989, 1991) has undertaken some useful explorations of this question, reflecting on how to articulate social with psychological findings. In another example, Wu and Martinson (1993) tested three different explanations of why lack of an intact family of orientation (a compositional variable) predisposes women to bear an out-of-wedlock child. They found little support for "socialization" or "social control" hypotheses, which reflect the culture and power constructs respectively. But they did find support for a "family instability and change" hypothesis. Since this entails merely an "event," as Wu and Martinson term it, it required further speculation to locate its theoretical efficacy. Wu and Martinson proposed that the event engendered "undesirable family circumstance," which was not further elaborated but which may imply a lack of emotional support (or low status-accord, as understood here).

Even if the proposed social structure model is successful in accounting for differences between social observables, it is still a black box, so to speak. Much will have to be done to illuminate the processes by which the social translates into the psychological.

An important body of work remains to be done to establish the relationships and directions of causal effect among the five proposed constructs in the social structural model. Indeed some of the sharpest arguments in sociological analysis bear on this question. Does structure precede culture or does culture precede structure (see Sewell 1992 for a review and a proposed solution to this problem). Do the constructs operate identically at both macro and micro levels (see Hechter 1983; Ritzer and Gindorf 1992)? How shall we accommodate the growing interest in agency, the sense that human actors actively make their fate with the seemingly deterministic notion of structure (Giddens 1984; Sewell 1992)? These and other questions are arrayed behind any proposal concerning social structure. Some of them are susceptible to empirical examination and others only to selection of a metatheoretical predisposition. In toto they provide a large agenda for future work.

Finally, there is a peculiar historical penumbra that shadows all sociological or social psychological conceptions of empirical social structure. For example, a few hundred years ago, one of the most important social observables was "blood," or nobility. Clearly, the difference between aristocrats and plebeians was important in its day. But that day has passed and we generally ignore that observable. As found earlier, the same may be occurring with respect to rural-urban residence and religion in the studies sampled. Indeed, the political thrust of our time is to reduce all social observables to indifference. Sex-gender, which is now so popular a topic for sociological and social psychological analysis, is on the agenda for elimination as a topic of interest as is race.

Indeed, all the so-called "isms" that pertain to social categories and are considered to reference prejudice or discrimination are targets for extinction. In a somewhat lurching, but mainly forward, movement, the U.S. Supreme Court has in recent years diminished the domain in which social structure puts members of given social categories at a disadvantage due solely to their membership in the category. Much remains to be done to eliminate category membership entirely as a definer of social structural position. However, should that task be completed, there will still be a social structure, interpretable as the division of labor, power and status relations, culture, composition and ecology. The difference will be that it will mean nothing at all social psychologically to be a member of certain observable social categories, for example, race, ethnicity, and sex-gender. Others, such as social class, would presumably retain their social psychological significance.

If we can, by mental experiment, think beyond the observable social categories that occupy substantial political and social psychological attention today, we would see simply individuals occupying positions in the social structure. Yet the social psychological effects of location in the structure would still be felt. For example, it would still matter social psychologically whether one were in a central or a peripheral position in the division of labor, except that one's position would not depend on any of the standard set of sociological observables, for example age, sex-gender, race, and so forth. My proposal here is that exposure to the division of labor, power and status, culture, composition, and ecology in given ways will always produce social psychological effects.

APPENDIX

References for Tables 1 and 2

Amato, P. R. 1990. "Personality and Social Network Involvement as Predictors of Helping Behavior in Everyday Life." *Social Psychology Quarterly* 53: 31-43.

Corsaro, W. A. 1992. "Interpretive Reproduction in Children's Peer Cultures." *Social Psychology Quarterly* 55: 160-77.

De Gilder, D. and H. A. M. Wilke. 1990. "Processing Sequential Status Information." *Social Psychology Quarterly* 53: 340-51.

Demo, D. H. and M. Hughes. 1990. "Socialization and Racial Identity among Black Americans." *Social Psychology Quarterly* 53: 364-74.

Feather, N. T. and I.R. McKee. 1993. "Global Self-Esteem and Attitudes Toward the High Achiever for Australian and Japanese Students." *Social Psychology Quarterly* 56: 65-76.

Felmlee, D., S. Sprecher, and E. Bassin. 1990. "The Dissolution of Intimate Relationships: A Hazard Model." *Social Psychology Quarterly* 53: 13-30.

Ford, R. and M. A. Blegen. 1992. "Offensive and Defensive Use of Punitive Tactics in Explicit Bargaining." *Social Psychology Quarterly* 55: 351-62.

Fox, M. Frank and V. C. Ferri. 1992. "Women, Men and Their Attributions for Success in Academe." *Social Psychology Quarterly* 55: 257-71.

Friedkin, N. E. 1990. "Social Networks in Structural Equation Models." *Social Psychology Quarterly* 53: 316-28.

Gecas, V. and M. A. Seff. 1990. "Social Class and Self-Esteem: Psychological Centrality, Compensation, and the Relative Effects of Work and Home." *Social Psychology Quarterly* 53: 165-73.

Grube, J. W. and M. Morgan. 1990. "Attitude-Social Support Interactions: Contingent Consistency Effects in the Prediction of Adolescent Smoking, Drinking, and Drug Use." *Social Psychology Quarterly* 53: 329-39.

Hallinan, M. T. and W. N. Kubitschek. 1990. "Sex and Race Effects of the Response to Intransitive Sentiment Relations." *Social Psychology Quarterly* 53: 252-63.

Hegtvedt, K. A. 1990. "The Effects of Relationship Structure on Emotional Responses to Inequity." *Social Psychology Quarterly* 53: 214-28.

Holtgraves, T. 1991. "Interpreting Questions and Replies: Effects of Face-Threat, Question Form, and Gender." *Social Psychology Quarterly* 54: 15-24.

Insko, C. A., J. Schopler, J. F. Kennedy, K. R. Dahl, K. A. Graetz, and S. M. Drigotas. 1992. "Individual-Group Discontinuity from the Differing Perspectives of Campbell's Realistic Group Conflict Theory and Tajfel and Turner's Social Identity Theory." *Social Psychology Quarterly* 55: 272-91.

Jackson, L. A. and L. A. Sullivan. 1990. "Perceptions of Multiple Role Participants." *Social Psychology Quarterly* 53: 274-82.

Jackson, P. B. 1992. "Specifying the Buffering Hypothesis: Support, Strain, and Depression." *Social Psychology Quarterly* 55: 363-78.

Kameda, T., M. F. Stasson, J. H. Davis, C. D. Parks, and S. K. Zimmerman. 1992. "Social Dilemmas, Subgroups, and Motivation Loss in Task-Oriented Groups: In Search of an 'Optimal' Team Size in Work Division." *Social Psychology Quarterly* 55: 47-56.

Kane, E. W. 1992. "Race, Gender, and Attitudes toward Gender Stratification." *Social Psychology Quarterly* 55: 311-20 .

Kemper, T. D. 1991. "Predicting Emotions from Social Relations." *Social Psychology Quarterly* 54: 330-42.

Komorita, S. S., K. F. Aquino, and A. L. Ellis. 1989. "Coalition Bargaining: A Comparison of Theories Based on Allocation Norms and Theories Based on Bargaining Strength." *Social Psychology Quarterly* 52: 183-96.

Kraus, L. A., M. H. Davis, D. Bazzini, M. Church, and C. M. Kirchman. 1993. "Personal and Social Influences on Loneliness: The Mediating Effect of Social Provisions." *Social Psychology Quarterly* 56: 37-53.

Lau, R. R. 1989. "Individual and Contextual Influences on Group Identification." *Social Psychology Quarterly* 52: 220-31.

Lerner, M. J. and P. R. Grant. 1990. "The Influences of Commitment to Justice and Ethnocentrism on Children's Allocation of Pay." *Social Psychology Quarterly* 53: 229-38.

Mazur, A., A. Booth, and J. M. Dabbs, Jr. 1992. "Testosterone and Chess Competition." *Social Psychology Quarterly* 55: 70-77.

Michaelson, A. and N. S. Contractor. 1992. "Structural Position and Perceived Similarity." *Social Psychology Quarterly* 55: 300-10.

Moen, P. and K. B. Forest. 1990. "Working Parents, Workplace Supports, and Well-Being: The Swedish Experience." *Social Psychology Quarterly* 53: 117-31.

Moore, D. 1991. "Entitlement and Justice Evaluations: Who Should Get More, and Why?" *Social Psychology Quarterly* 54: 208-23.

Mortimer, J. T. and J. Lorence. 1989. "Satisfaction and Involvement: Disentangling a Deceptively Simple Relationship." *Social Psychology Quarterly* 52: 249-65.

Naoi, M. and C. Schooler. 1990. "Psychological Consequences of Occupational Conditions Among Japanese Wives." *Social Psychology Quarterly* 53: 100-16.

Parcel, T. L. and E. G. Menaghan. 1990. "Maternal Working Conditions and Children's Verbal Facility: Studying the Intergenerational Transmission of Inequality from Mothers to Young Children." *Social Psychology Quarterly* 53: 132-47.

Rabow, J., M. D. Newcomb, M. A. Monto, and A. C. R. Hernandez. 1990. "Altruism in Drunk Driving Situations: Personal and Situational Factors in Intervention." *Social Psychology Quarterly* 53: 199-213.

Riches, P. and M. Foddy. 1989. "Ethnic Accent as a Status Cue." *Social Psychology Quarterly* 52: 197-206.

Ross, C. E. and J. Mirowsky. 1992. "Households, Employment, and the Sense of Control." *Social Psychology Quarterly* 55: 217-35.

Sarup, G., R. W. Suchner, and G. Gaylord. 1991. "Contrast Effects and Attitude Change: A Test of the Two-Stage Hypothesis of Social Judgment Theory." *Social Psychology Quarterly* 54: 364-72.

Scanzoni, J. and D. D. Godwin. 1990. "Negotiation Effectiveness and Acceptable Outcomes." *Social Psychology Quarterly* 53: 239-51.

Schuman, H., L. Bobo, and M. Krysan. 1992. "Authoritarianism in the General Population: The Education Interaction Hypothesis." *Social Psychology Quarterly* 55: 379-87.

Schwalbe, M. L. and C. L. Staples. 1991. "Gender Differences in Sources of Self-Esteem." *Social Psychology Quarterly* 54: 158-68.

Schwartz, S. H. and N. Struch. 1990. "Values and Intergroup Social Motives: A Study of Israeli and German Students." *Social Psychology Quarterly* 53: 185-98.

Scott, J. 1989. "Conflicting Beliefs About Abortion: Legal Approval and Moral Doubts." *Social Psychology Quarterly* 52: 319-26.

Sev'er, A. 1989. "Simultaneous Effects of Status and Task Cues: Combining, Eliminating, or Buffering?" *Social Psychology Quarterly* 52: 327-35.

Shanahan, M. J., M. Finch, J. T. Mortimer, and S. Ryu. 1991. "Adolescent Work Experience and Depressive Affect." *Social Psychology Quarterly* 54: 299-317.

Simon, R. W., D. Eder, and C. Evans. 1992. "The Development of Feeling Norms Underlying Romantic Love Among Adolescent Females." *Social Psychology Quarterly* 55: 29-46.

Skvoretz, J. and D. Willer. 1991. "Power in Exchange Networks: Setting and Structural Variations." *Social Psychology Quarterly* 54: 224-38.

Smith, T. E. 1990. "Academic Achievement and Teaching Younger Siblings." *Social Psychology Quarterly* 53: 352-63.

———. 1993. "Growth in Academic Achievement and Teaching Younger Siblings." *Social Psychology Quarterly* 56: 77-85.

Sprecher, S. 1992. "How Men and Women Expect to Feel and Behave in Response to Inequity in Close Relationships." *Social Psychology Quarterly* 55: 57-69.

Stasson, M. F., T. Kameda, C. D. Parks, S. K.Zimmerman, and J. H. Davis. 1991. "Effects of Assigned Group Consensus Requirement on Group Problem Solving and Group Members' Learning." *Social Psychology Quarterly* 54: 25-35.

Stevens, G., D. Owens, and E. Schaefer. 1990. "Education and Attractiveness in Marriage Choices." *Social Psychology Quarterly* 53: 62-70.

Stewart, P. A. and J. C. Moore, Jr. 1992. "Wage Disparities and Performance Expectations." *Social Psychology Quarterly* 55: 78-85.

Thoits, P. A. 1991. "On Merging Identity Theory and Stress Research." *Social Psychology Quarterly* 54: 101-12.

———. 1992. "Identity Structures and Psychological Well-Being: Gender and Marital Status Comparisons." *Social Psychology Quarterly* 55: 236-56.

Verkuyten, M. 1991. "Self-Definition and Ingroup Formation Among Ethnic Minorities in the Netherlands. *Social Psychology Quarterly* 54: 280-86.

Whitbeck, L. B., R. L. Simons, R. D. Conger, F. O.Lorenz, S. Huck, and G. H. Elder, Jr. 1991. "Family Economic Hardship, Parental Support and Adolescent Self-Esteem." *Social Psychology Quarterly* 54: 353-63.

Wiltfang, G. L. and M. Scarbecz. 1990. "Social Class and Adolescent Self-Esteem: Another Look."
    Social Psychology Quarterly 53: 174-82.
Young, R. L. 1991. "Race, Conceptions of Crime and Justice, and Support for the Death Penalty."
    Social Psychology Quarterly 54: 67-75.

NOTES

1.   I employ the term sociological social psychology to mean the type of analysis in which
social variables (of whatever kind) are independent and psychological or individual level variables
are dependent. This usage has longstanding currency among sociologist, for example Schnore
(1961), although it differs from House's (1977) denomination of social psychology into "three
faces." What I call sociological social psychology incorporates both his "psychological social
psychology" and "psychological sociology." I believe it is reasonable to supplant House's tripartite
distinction because it was merely historical, that is, based on prevailing practices rather than an
analytical or theoretically meaningful division.

2.   Although these descriptive variables appear less often in laboratory-experimental studies
of group process, they are present there too and require better explication as theoretical constructs.

3.   Notwithstanding the merits of the proposal, it has failed of widespread adoption at least
in part because of the contemporaneous emergence of status politics. The observables of race,
sex-gender, ethnicity, and the like, have taken on a broad significance with respect to obtaining
rights previously denied. To swallow these categories by constructs of no resonance for political
or social mobilization purposes would diminish their political effectiveness. Mildly ironic, however,
is that the exponents of rights for groups that are denominated by observables do so in the name
of theoretical constructs, for example, *power* and *status*.

4.   For purposes of understanding social structure, the terms sociological and social psychological
are used interchangeably since in this context they have identical meaning, significance, and usage.

5.   Although it would suit sociologists' imperial ambitions very nicely, not all differences
between members of socially-observable categories are to be explained socially. Certainly, some
differences between the sexes are biological. Some physically aggressive behavior and other forms
of deviance are at least partially explained endocrinologically (Dabbs, Frady, Carr, and Besch
1987; Dabbs and Morris 1990; Booth and Osgood 1993), as are differences in sexual behavior
by adolescent males (Udry 1988). Recent advances in genetics and the understanding of other
biological processes have provided a stronger foundation of credibility for those advancing
biological explanations of descriptive differences between social categories than we have seen for
some time. Sociobiology is a somewhat threatening product of such efforts.

6.   The term variable is applied here in only a rough sense, since some of the variables are
actually *supervariables* that can be understood in a variety of aspects thus leading to many
subvariables that are useful held under the same rubric. For example, one determinant of job
performance may be mode of transportation to work, for example, private car, commuter train,
subway, bus, taxi, walking, and so forth. Each of these types of transportation implies multiple
considerations, such as seating comfort, noise level, amount of light, smoothness of ride, carbon
monoxide level, social density, and so on. The supervariable, mode of transportation, thus implies
a multiplicity of subvariables.

7.   This is not to say that once the division of labor has been found to "explain" differences
in some dependent variable that the work has ended. For example, when we find that centrality
in the division for labor is often associated with leadership we must ask what properties of centrality
produce this outcome. But this question could not be addressed until the prior question was
answered, namely that it is something about the division of labor that is related to leadership.

8. These do not necessarily exhaust categories of the division of labor. Like all constructs, the division of labor is open to revision, expansion, and refinement.

9. Since Marx saw virtually all labor devolving to the level of factory production the specialization of factory labor was the bane he railed against. He had not figured on the enormous elaboration of symbolic content in professions and many white-collar positions, which, though specialized, do not lead to the kind of alienation Marx assigned to specialization per se.

10. The power and status social structure also obtains between individuals in the interpersonal sense in addition to what prevails between members of social categories. The same is true of the division of labor previously discussed.

11. Categorical observables, such as age, sex-gender, and so forth are frequently presented, or examined, first in terms of their compositional and or ecological aspects, for example, the proportions of males and females in a group or their location in different parts of a social network. Normally, the next step is to examine the distribution of observables for their construct significance, how the distribution is related to the division of labor, power and status relations, and culture.

12. Although composition and ecology are observables, they are group properties, hence are exterior and constrain individuals. They also are structural conditions as far as individuals are concerned, hence I will treat them as constructs, keeping in mind that they obtain their effects through the division of labor, power and status relations, and culture.

13. Although I have referred to the set of constructs proposed here as a model, it is not a model in the technical sense that it leads to specific testable propositions about empirical phenomena. Thus, is more accurate to designate the set of constructs a a metatheory, analytic framework, or orienting strategy (Wagner 1984). Notwithstanding this point of terminological niceness, I will continue to use the term model because it is short and widely used to denote even metatheories and the like.

14. Some journals devoted to social psychology may outrank *Social Psychology Quarterly* in prestige, for example, *Journal of Personality and Social Psychology,* but the former claims sociological precedence because it is an official publication of the American Sociological Association.

15. Not all articles were suitable for the analysis conducted here. Omitted were articles devoted to methods, to the attitude-behavior issue, to self or identity where there was no reference to recognizable groups or theoretical constructs of a social nature, review articles that did not contain original work, and articles introducing special issues, the Cooley-Mead award winner, and the Cooley-Mead Award winner's address.

16. References cited in Tables 1 and 2 are contained in the appendix.

17. On the other hand, neighborhood, taken as a residence variable, is important in many types of urban research dealing with crime, drug use, housing segregation, and the like. The sample is inadequate in that respect.

18. Another possible explanation is the availability of the journal *Environmental Psychology* which is specifically devoted to ecological studies, hence may simply drain them away from other publications.

REFERENCES

Amato, P. 1990. "Personality and Social Network Involvement as Predictors of Helping Behavior in Everyday Life." *Social Psychology Quarterly* 53: 31-43.

Asch, S. 1956. "Studies of Independence and Submission to Group Pressure: I. A Minority of One Against a Unanimous Majority." *Psychological Monographs* 70(9): 1-70.

Bavelas, A. 1948. "A Mathematical Model for Group Structures." *Applied Anthropology* 7: 16-30.

Berger, J. A., M. H. Fisek, R. Z. Norman, and M. Zelditch, Jr. 1977. *Status Characteristics and Social Interaction: An Expectation States Approach.* New York: Elsevier Scientific.

Berger, J. A., M. H. Fisek, R. Z. Norman, and D. G. Wagner. 1985. "Formation of Reward Expectations in Status Situations." Pp. 215-61 in *Status, Rewards and Influence*, edited by J. Berger and M. Zelditch, Jr. San Francisco: Jossey-Bass.

Berkowitz, L. 1956. "Personality and Group Position." *Sociometry* 19: 210-22.

Blau, P. M. 1975. "Structural Constraints of Status Complements." Pp. 117-38 in *The Idea of Social Structure: Papers in Honor of Robert K. Merton*, edited by L. M. Coser. New York: Harcourt Brace Jovanovich.

Blauner, R. 1964. *Alienation and Freedom: The Factory Worker and His Industry.* Chicago: University of Chicago Press.

Blumer, H. 1937. "Social Disorganization and Individual Disorganization." *American Journal of Sociology* 42: 871-77.

Booth, A. and D. W. Osgood. 1993. "The Influence of Testosterone on Deviance in Adulthood: Assessing and Explaining the Relationship." *Criminology* 31: 93-117.

Brim, O. G., Jr. 1958. "Family Structure and Sex RoleLearning by Children: A Further Analysis of Helen Koch's Data." *Sociometry* 21: 1-16.

Burt, R. S. 1992. *Structural Holes: The Social Structure of Competition.* Cambridge: Harvard University Press.

Campbell, D. T. and D. W. Fiske. 1959. "Convergent and Discriminant Validation by Multitrait-Multimethod Matrix." *Psychological Bulletin* 56: 81-105.

Cattell, R. 1973. *Personality and Mood by Questionnaire.* London: Methuen.

Collins, R. 1975. *Conflict Sociology.* New York: Academic.

Cook, K. S. and R. Emerson. 1978. "Power, Equity and Commitment in Exchange Networks." *American Sociological Review* 43: 721- 39.

Cook, K. S., K. A. Hegtvedt, and T. Yamagishi. 1987. "Structural Inequality, Legitimation and Reactions to Inequity in Exchange Networks." Pp. 291-308 in *Status Generalization: New Theory and Research*, edited by M. Webster and M. Foschi. Stanford: Stanford University Press.

Corsaro, W. 1992. "Interpretive Reproduction in Children's Peer Cultures." *Social Psychology Quarterly* 55: 160-77.

Dabbs, J., Jr., R. Frady, T. Carr, and N. Besch. 1987. "Saliva Testosterone and Criminal Violence in Young Adult Prison Inmates." *Psychosomatic Medicine* 49:174-182.

Dabbs, J., Jr. and R. Morris. 1990. "Testosterone, Social Class, and Antisocial Behavior in a Sample of 4,462 Men." *Psychological Science* 1: 209-11.

Durkheim, E. 1933. *The Division of Labor in Society.* Translated by G. Simpson. New York: Macmillan.

————. 1938. *The Rules of Sociological Method.* Translated by S. A. Solvay and J. H. Mueller. Chicago: University of Chicago Press.

Easterlin, R. 1980. *Birth and Fortune: The Impact of Numbers on Personal Welfare.* New York: Basic.

Emerson, R. 1962. "Power-Dependence Relations." *American Sociological Review* 40: 252-57.

Festinger, L., S. Schachter, and K. W. Back. 1950. *Social Pressures in Informal Groups.* Stanford, CA: Stanford University Press.

Garfinkel, I. and S. S. McLanahan. 1986. *Single Mothers and their Children: A New American Dilemma.* Washington, DC: Urban Institute.

Gecas, V. and M. A. Seff. 1990. "Social Class and Self-Esteem: Psychological Centrality, Compensation and the Relative Effects of Work and Home." *Social Psychology Quarterly* 53: 165-73.

Giddens, A. 1984. *The Constitution of Society: Outline of the Theory of Structuration.* Berkeley: University of California Press.

Guttentag, M. and P. F. Secord. 1986. *Too Many Women: The Sex Ratio Question.* Beverley Hills: Sage.

Hall, E. T. 1959. *The Silent Language.* New York: Doubleday.

Hawley, A. 1950. *Human Ecology.* New York: Ronald.

Hechter, M. 1983. *The Microfoundations of Macrosociology.* Philadelphia: Temple University Press.

Heise, D. 1979. *Understanding Events: Affect and the Construction of Social Action.* New York: Cambridge University Press.

House, J. S. 1977. "The Three Faces of Social Psychology." *Sociometry* 40: 161-77.

Insko, C. A., R. L. Pinkley, K. Haring, and B. Holton. 1987. "Minimal Conditions for Real Groups: Mere Categorization or Competitive Between Category Behavior." *Representative Research in Social Psychology* 17: 5-36.

Kameda, T., M. F. Stasson, J. H. Davis, C. D. Parks, and S. K. Zimmerman. 1992. "Social Dilemmas, Subgroups, and Motivation Loss in Task-Oriented Groups: In Search of an 'Optimal' Team Size in Work Division." *Social Psychology Quarterly* 53: 47-56.

Kemper, T. D. 1972. "The Division of Labor: A Post-Durkheimian Analytical View." *American Sociological Review* 37: 739-53.

_____ . 1975. "Emile Durkheim and the Division of Labor." *Sociological Quarterly* 16: 190-206.

_____ . 1978. *A Social Interactional Theory of Emotions.* New York: Wiley.

_____ . 1991. "Predicting Emotions from Social Relations." *Social Psychology Quarterly* 54: 330-42.

Kemper, T. D. and R. Collins. 1990. "Dimensions of Microinteraction." *American Journal of Sociology* 96: 32-68.

Kohn, M. 1977. *Class and Conformity: A Study in Values* (2nd edition). Chicago: University of Chicago Press.

_____ . 1983. *Work and Personality: An Inquiry into the Impact of Social Stratification.* Norwood, NJ: Ablex.

Kohn, M. and C. Schooler. 1982. "Job Conditions and Personality: A Longitudinal Study of their Reciprocal Effects." *American Journal of Sociology* 87: 1257-86.

Kraus, L. A., M. H. Davis, D. Bazzini, M. Church, and C. M. Kirchman. 1993. "Personal and Social Influences on Loneliness: The Mediating Effect of Social Provisions." *Social Psychology Quarterly* 56: 37-53.

Krug, S. F. and R. W. Kulhavy. 1973. "Personality Differences across Regions of the United States." *Journal of Social Psychology* 91: 73-79.

Lau, R. 1989. "Individual and Group Contextual Influences on Group Identification." *Social Psychology Quarterly* 52: 220-31.

Leach, R. 1954. *Political Systems of Highland Burma.* Boston: Beacon.

Leavitt, H. 1951. "Some Effects of Certain Communication Patterns on Group Performance." *Journal of Abnormal and Social Psychology* 45: 38-50.

Lepore, S. J., G. Evans, and M. N. Palsane. 1991. "Social Hassles and Psychological Health in the context of Chronic Crowding." *Journal of Health and Social Behavior* 32: 357-67.

Lepore, S. J., G. W. Evans, and M. L. Schneider. 1992. "Role of Control and Social Support in Explaining the Stress of Hassles and Crowding." *Environment and Behavior* 24: 795-811.

Lipset, S. M. 1960. *Political Man.* New York: Doubleday.

Mann, M. 1970. "The Social Cohesion of Liberal Democracy." *American Sociological Review* 35: 423-39.

Markovsky, B., D. Willer, and T. Patton. 1988. "Power Relations in Exchange Networks." *American Sociological Review* 53: 220-36.

Markovsky, B., J. Skvoretz, D. Willer, M. J. Lovaglia, and J. Erger. 1993. "The Seeds of Weak Power: An Extension of Network Exchange Theory." *American Sociological Review* 58: 197-209.

Marx, K. 1964. *Selected Writings in Sociology and Social Philosophy.* Translated by T. B. Bottomore and edited by T. B. Bottomore and M. Rubel. New York: McGraw-Hill.

Mayhew, L. 1973. "System Size and Ruling Elites." *American Sociological Review* 38: 468-475.

Mead, G. H. 1934. *Mind, Self and Society.* Chicago: University of Chicago Press.

Molm, L. 1985. "Relative Effects of Individual Dependencies: Further Tests of the Relation between Power Imbalance and Power Use." *Social Forces* 63: 810-37.

―――――. 1987. "Linking Power Structure and Power Use." Pp. 101-29 in *Social Exchange Theory*, edited by K. S. Cook. Newbury Park, CA: Sage.

Nadel, S. F. 1964. *The Theory of Social Structure.* New York: Free Press.

Pallas, A., D. R. Entwisle, K. L. Alexander, and P. Weinstein. 1990. "Social Structure and the Development of Self-Esteem in Young Children." *Social Psychology Quarterly* 53: 302-15.

Parcel, T. L. and E. G. Menaghan. 1990. "Maternal Working Conditions and Children's Verbal Facility: Studying the Intergenerational Transmission of Inequality from Mothers to Young Children." *Social Psychology Quarterly* 53: 132-47.

Parson, T. 1951. *The Social System.* Glencoe, IL: Free Press.

―――――. 1967. *Sociological Theory and Modern Society.* New York: Free Press.

Riches, P. and M. Foddy. 1989. "Ethnic Accent as a Status Cue." *Social Psychology Quarterly* 52: 197-206.

Ridgeway, C. and J. Berger. 1988. "The Legitimation of Power and Prestige Orders in Task Groups." Pp. 207-31 in *Status Generalizations: New Theory and Research*, edited by M. Webster, Jr. and M. Foschi. Stanford, CA: Stanford University Press.

Ritzer, G. and P. Gindoff. 1992. "Methodological Relationalism: Lessons for and from Social Psychology." *Social Psychology Quarterly* 55: 128-40.

Rosenberg, M. 1962. "The Dissonant Religious Context and Emotional Disturbance." *American Journal of Sociology* 68: 1-10.

―――――. 1990. "Reflexivity and Emotions." *Social Psychology Quarterly* 53: 3-12.

Ruback, B. R. and J. Pandey. 1992. "Very Hot and Very Crowded: Quasi-Experimental Investigations of Indian 'Tempos'." *Environment and Behavior* 24: 527-54.

Scanzoni, J. and D. D. Godwin. 1990. "Negotiation Effectiveness and Acceptable Outcomes." *Social Psychology Quarterly* 53: 239-51.

Schnore, L. F. 1961. "The Myth of Human Ecology." *Sociological Inquiry* 31: 128-39.

Schooler, C. 1987. "Psychological Effects of Complex Environments during the Life Span: A Review and Theory." Pp. 24-49 in *Cognitive Functioning and Social Structures over the Life Course*, edited by C. Schooler and K. W. Shaie. Norwood, NJ: Ablex.

―――――. 1989. "Social Structure Effects and Experimental Situations: Mutual Lessons of Cognitive and Social Science." Pp. 129-47 in *Social Structure and Aging: Psychological Processes*, edited by K. W. Schaie and C. Schooler. Hillsdale, NJ: Erlbaum.

―――――. 1991. "Interdisciplinary Lessons: The Two Social Psychologies from the Perspective of a Psychologist Practicing Sociology." Pp. 71-81 in *The Future of Social Psychology:*

Defining the Relationship between Sociology and Psychology, edited by C. W. Stephan, W. G. Stephan, and T. F. Pettigrew. New York: Springer-Verlag.

Seeman, M. 1959. "On the Meaning of Alienation." *American Sociological Review* 24: 783-91.

Sewell, W. H., Jr. 1992. "A Theory of Structure: Duality, Agency, and Transformation." *American Journal of Sociology* 98: 1-29.

Smelser, N. J. 1959. *Social Change in the Industrial Revolution.* Chicago: University of Chicago Press.

Smith-Lovin, L. and D. Heise. 1988. *Analyzing Social Interaction: Advances in Affect Control Theory.* New York: Gordon and Breach.

Sommers, R. 1959. "Studies in Personal Space." *Sociometry* 22: 247-60.

Stanton, A. H. and M. S. Schwartz. 1954. *The Mental Hospital.* New York: Basic.

Strodtbeck, F. L. and H. Hook. 1961. "The Social Dimensionsof a 12-Man Jury Table." *Sociometry* 24: 397-415.

Turner, B. S. and N. Abercrombie. 1980. *The Dominant Ideology Thesis.* London: G. Allen and Unwin.

Udry, R. 1988. "Biological Predispositions and Social Control in Adolescent Sexual Behavior." *American Sociological Review* 53: 709-22.

Wagner, D. 1984. *The Growth of Sociological Theories.* Beverly Hills: Sage.

Warner, R. S. 1993. "Work in Progress toward and New Paradigm for the Sociological Study of Religion in the United States." *American Journal of Sociology* 98: 1044-93.

Weber, M. 1946. *From Max Weber: Essays in Sociology.* Translated and edited by H. H. Gerth and C. Wright Mills. New York: Oxford University Press.

_____. 1947. *The Theory of Social and Economic Organization.* Translated by A. M. Henderson and T. Parsons. New York: Oxford University Press.

Whitbeck, L. B., R. L. Simons, R. D. Conger, F. O. Lorenz, S. Huck, and G. H. Elder, Jr. 1991. "Family Economic Hardship, Parental Support and Adolescent Self-Esteem. *Social Psychology Quarterly* 54: 353-63.

White, J. D., A. Tashchian, and R. Ohanian. 1991. "An Exploration into the Scale of Consumer Confidence: Dimensions, Antecedents, and Consequences." *Journal of Social Behavior and Personality* 6: 509-28.

Willer, D. 1987. *Theory and Experimental Investigation of Social Structures.* New York: Gordon and Breach.

Willer, D. and M. Webster, Jr. 1970. "Theoretical Concepts and Observables." *American Sociological Review* 35: 748-57.

Wu, L. L. and B. C. Martinson. 1993. "Family Structure and the Risk of a Premarital Birth." *American Sociological Review* 58: 210-32.

SEXUAL ORIENTATION AS A DIFFUSE STATUS CHARACTERISTIC:

IMPLICATIONS FOR SMALL GROUP INTERACTION

Cathryn Johnson

ABSTRACT

In this chapter I argue that sexual orientation is a diffuse status characteristic in our society. I demonstrate this by examining (1) the negative stereotypes of and attitudes toward gay men and lesbians, (2) the prevalence of anti-lesbian/gay interpersonal and institutional discrimination, and (3) preliminary evidence that demonstrates an association between sexual orientation and general expectation states. Given that states of sexual orientation, namely straight and lesbian/gay, are differentially valued in society, what are the implications for interaction in task groups? To address this, I examine the role of indicative categorical cues and expressive cues in the activation of sexual orientation in task groups. The most interesting consequence of the activation of sexual orientation through expressive cues is the possibility of the simultaneous activation of a second characteristic, gender, in same-sex dyads. Based on this possibility, I provide some preliminary predictions of the effects of sexual orientation on the development of performance expectations and, ultimately, behavior in same-sex dyads.

Advances in Group Processes, Volume 12, pages 115-137.
ISBN: 1-55938-872-2

According to status characteristics theory, diffuse status characteristics, such as gender, race, and age, are defined as having at least two states that are differentially evaluated in society in terms of esteem and value and are associated with distinct sets of specific and general expectation states (Berger, Conner, and Fisek 1974; Berger, Fisek, Norman, and Zelditch 1977). For example, gender is a diffuse status characteristic in that it has two states, male and female, which are differentially valued, males being more highly valued than females. Berger, Rosenholtz, and Zelditch (1980) argued that gender is a status characteristic in our society for at least three reasons. First, there is a more favorable overall evaluation of males than females, evidenced by the fact that adjectives associated with females are rated less favorably than adjectives associated with males. Second, there is a high level of agreement among males and females on the traits that differentiate male from female. Third, there is a larger number of positive traits attributed to males than to females (p. 494). In addition, given that male and female are differentially valued, gender carries with it general expectations for performance that are incorporated into small task groups whenever gender is activated in the situation. Being male or female, then, is culturally associated with expectations for superior (or inferior) ability.

In this paper, I argue that sexual orientation is also a diffuse status characteristic in our society and, therefore, has implications for social interaction in small informal task groups. My task is to demonstrate that this characteristic currently has status value in our society, where *straight* is the high state and *lesbian/gay* is the low state, to describe some consequences of this fact, and then to provide preliminary predictions on how the activation of sexual orientation may affect interaction in task groups under specific conditions.[1] As will be shown, sexual orientation is an interesting characteristic theoretically for several reasons. First, it is, in many situations, a hidden characteristic, often inferred, sometimes correctly and sometimes incorrectly, by appearance and other verbal and nonverbal behavior. Second, it is intricately related to perceived deviation from gender roles (gender role nonconformity) in our society. As a result, it is a status characteristic that is closely linked to another status characteristic, gender status. Third, there are implications for the activation of this characteristic in face-to-face interaction, several of them related to other status-related behaviors previously studied. Fourth, viewing sexual orientation this way suggests interaction techniques developed for other status characteristics may be adapted to overcome undesirable features of status generalization based on sexual orientation.

To achieve my tasks, I divide this paper into four sections. In the first section, I present a brief summary of status characteristics theory, with particular focus on an important extension of this theory involving the relationship of task and categorical cues to status in small group interaction (Ridgeway, Berger, and Smith 1985; Berger, Webster, Ridgeway, and Rosenholtz 1986). Work on status cues is crucial to my argument about how sexual orientation is activated and how it becomes relevant to the development of status hierarchies in small task groups. In the second section, I show how sexual orientation is a diffuse status characteristic in our society by first providing data on negative stereotypical traits that are associated with gay men and lesbians. One consequence of these negative stereotypes is anti- lesbian/ gay discrimination, both interpersonal and institutional. I then present results from a recent study which demonstrate a link between sexual orientation and general expectation states. Finally, I examine the relationship between sexual orientation and perceived deviation from gender roles by looking at the relationship between gay male and lesbian stereotypes and gender stereotypes. After establishing that sexual orientation is a diffuse status characteristic, in the third section I examine how sexual orientation is activated in task groups through the use of categorical cues. I also show how gender may be activated under certain conditions, given the relationship between sexual orientation and gender status. Finally, in the last section, I offer several preliminary predictions, under certain conditions, for interaction in same-sex dyads where members differentiate on sexual orientation, discuss two issues critical to the empirical testing of these predictions, and present an outline of an experimental study that could be used to appropriately test these predictions.

STATUS CHARACTERISTICS THEORY AND STATUS CUES

In general, status characteristics theory provides an explanation for the development and maintenance of status hierarchies in small task groups where group members are oriented toward a collective goal or task, for example, discussion groups or decision-making groups. Specifically, it argues that diffuse status characteristics carry with them specific and general performance expectations that are implicitly assumed to apply to a wide range of situations (Ridgeway, Berger, and Smith 1985). A performance expectation is an anticipation of one's own and the other's ability to make useful contributions to the task. Activation of status characteristics occur either when members are differentiated on states of characteristics (as in mixed-sex or mixed-race groups) or when the task is related to the status characteristic (as in gender-stereotyped tasks). When a diffuse status characteristic differentiates members, the cultural expectations associated with the characteristic are activated, and this sets the foundation for the expectations members form for their own and each others' performance at the task, even when the characteristic is objectively irrelevant to the task (as sexual orientation and gender often are).

The theory predicts that, for example, in informal task groups where members are equal on all other status characteristics (e.g., age, education, and race), if gender differentiates them, both men and women will expect men to perform better at the task, unless there is convincing evidence that gender is not relevant to the task in this case (Pugh and Wahrman 1985). Because both men and women in mixed-sex groups have higher performance expectations for men, men will be likely to contribute more to the task, have more opportunities to contribute, receive more positive evaluations for their contributions, and be more influential in group decisions than women (Berger, Conner, and Fisek 1974; Berger, Fisek, Norman, and Zelditch 1977; Berger, Webster, Ridgeway, and Rosenholtz 1986). There is solid empirical evidence to support this formulation of the theory (Strodtbeck and Mann 1956; Lockheed and Hall 1976; Piliavin and Martin; 1978; Wood and Karten 1986).

Status cues are indicators or identifiers of the different social statuses and task abilities people possess. They are observable aspects of appearance, behavior, or surrounding possessions used by group members to make inferences about each others' status. Ridgeway, Berger, and Smith (1985) and Berger, Webster, Ridgeway, and Rosenholtz (1986) extended status characteristics theory by examining the relationship between different types of status cues and status in task groups. Berger, Webster, Ridgeway, and Rosenholtz (1986) offer two dimensions of status cues in their typology. First, they differentiate task cues and categorical cues. Task cues are nonverbal or verbal behaviors or signs that give information about performances taking place in the immediate interaction situation (Ridgeway, Berger, and Smith 1985, p. 963). Members use this information to infer how well they all will perform and are performing at the group task. Examples of task cues are response latency, eye gaze, verbal loudness, fluency, body posture, as well as statements that directly refer to a member's ability such as, "I have had a great deal of experience with this type of problem," or, "I have the ability in general to solve problems" (Berger, Webster, Ridgeway, and Rosenholtz 1986, p. 6).

Categorical cues refer to aspects of a person's appearance, behavior, or possessions that give information on the social groups to which she or he belongs in the larger society. They are used to identify states of status characteristics possessed by group members and to activate beliefs and stereotypes associated with these characteristics (Berger, Webster, Ridgeway, and Rosenholtz 1986, p. 13). Examples of categorical cues are skin color, accent, word usage, diplomas on the wall, or declarations of social position outside the group, such as, "I am a mathematician" (Berger, Webster, Ridgeway, and Rosenholtz 1986, p. 6).

In addition, task and categorical cues are further divided into indicative or expressive cues. Indicative cues make clear direct claims to possessing a particular state of a status class (indicative categorical cues) or to possessing states of abilities and competence (indicative task cues). Indicative categorical cues are such things as diplomas or licenses and statements such as "I am a Harvard Ph.D." Indicative task cues are statements such as "I don't know anything about this kind of work" (Berger, Webster, Ridgeway, and Rosenholtz 1986, p. 4).

Expressive cues are behaviors or signs "given off" (Goffman 1959) during interaction rather than directly presented. These cues provide information about the diffuse status groups to which a member belongs (expressive categorical cues) or information about abilities of the group member (expressive task cues). Examples of the former include accent, dress, style of speech, body movements, and skin color. Examples of the latter are posture, type and duration of eye contact, speech speed, speech fluency and hesitancy, and maintaining a minority position (Berger, Webster, Ridgeway, and Rosenholtz 1986, p. 6).

Categorical cues, both indicative and expressive, seem to play a major role in the process by which diffuse status characteristics become salient to actors in the beginning of their interaction. They may be used to identify these status characteristics and each members' relation to them (Ridgeway, Berger, and Smith 1985).

In the extension of status characteristics theory, the focus has been on the relationship between task cues and status, both in peer groups where groups do not initially differentiate on status characteristics and in status unequal groups where members do differentiate. Ridgeway, Berger, and Smith (1985) and Berger, Webster, Ridgeway, and Rosenholtz (1986) argue that, in peer groups, differences in task cue levels give rise to differential performance expectations. Certain task cue behaviors (e.g., speaking fluently with a firm tone of voice and using direct eye contact) lead to attributions of competence and, therefore, to high expectations for those members displaying them, while other task cue behaviors lead to low expectations. These performance expectations, in turn, affect the development of the status hierarchy in the group, where members with high expectations will talk more, be listened to more often, and will be more influential. They will also continue to use those task cues associated with a high level of competence, while members with low expectations will continue to display those task cues associated with a low level of competence, thereby maintaining the status hierarchy.

In a group of initial status unequals where members differentiate on status characteristics, the causal relationship between task behaviors and status is reversed. Once these characteristics become salient, they become the basis for the difference in performance expectations that a member holds for herself and for any other interactant. These differences in performance expectations then are expressed in differential levels of task cues (Berger, Webster, Ridgeway, and Rosenholtz 1986). Differences in task cues are read by others as differences in competence. High levels of task cues, then, express high performance expectations and they also serve to maintain and justify them. This causal argument about the relation between task cues and status has been empirically tested and supported in groups of unequal status (Ridgeway, Berger, and Smith 1985).

SEXUAL ORIENTATION AS A DIFFUSE STATUS CHARACTERISTIC

For something to be a diffuse status characteristic its states must be differentially valued in the larger society, where one state is evaluated as inferior to another, and these states must be associated with distinct sets of specific and general expectation states (Berger, Rosenholtz, and Zelditch 1980). To demonstrate that states of sexual orientation are differentially valued in our society, I examine general stereotypes of and attitudes toward gays and lesbians and provide a brief discussion of the prevalence of antigay/lesbian discrimination.

I regard sexual orientation as a nominal characteristic since we tend to categorize individuals as straight, gay male, lesbian, or bisexual, rather than perceive it as a graduated or ordinal characteristic (Ridgeway 1991). I focus on two categories or states of sexual orientation—straight and lesbian/gay— for simplicity at this stage. Bisexuality is more complex, given that those who claim bisexuality are often viewed with suspicion from both the straight world and the gay community (Garnets and Kimmel 1993).

Gay and Lesbian Stereotypes

There are two distinct sets of stereotypes involved when considering homosexuality: (1) the stereotypes of "homosexuals" versus "heterosexuals" and (2) the stereotypes of gay men and lesbians separately (Hayworth 1991). As we shall see, the second type is intertwined with gender stereotypes.

Concerning the first set, there is a consensus in society among many people, if not most, that it is "better" and more worthy to be straight than gay. According to public opinion polls in the 1970s and 1980s, about two-thirds of U.S. citizens thought that homosexuality is morally wrong or undesirable and more than half felt that homosexual activity between consenting adults should be illegal (Herek 1991). In addition, homosexuals have long labored under the perception of being mentally ill. Although the American Psychiatric Association no longer considers "homosexuality" as a mental illness, it is still the consensual belief in this society that "homosexuality" is abnormal, while "heterosexuality" is considered normal (Garnets and Kimmel 1993).

As is problematic with any kind of stereotype, "homosexuals" are often viewed as a one-dimensional homogeneous group of people (usually men—women are often forgotten completely) (Herdt 1989). In addition, to be called a "homosexual" is to be degraded and devalued in our society (Plummer 1975; Preston and Stanley 1987) and most "homosexual" characteristics are held in low esteem by most segments of society (Herek 1991).

What are these stereotypic characteristics or traits? In order to understand them, the stereotypes of gay men and lesbians should be examined separately. It seems that much, although not all, of the content of antigay/lesbian stereotypes is tied to gender nonconformity (Herek 1991, 1993). Both men and women who manifest characteristics inconsistent with those culturally prescribed are more likely to be labeled as "homosexual," even when they are not (Deaux and Lewis 1984; Herek 1984; Blumenfeld 1992; Morgan and Brown 1993). Individuals who do not fit into the assigned gender roles may and often do become targets of verbal or physical abuse (Blumenfeld 1992; Elze 1992; Pellegrini 1992). Homosexuality, then, is commonly associated with deviation from our culturally defined masculinity and femininity; it is equated with violating norms of gender. Heterosexuality, in contrast, is associated with "normal" masculinity and "normal" femininity (Herek 1993; Woods 1993). The complexity of the relationship between sexual orientation and perceptions of gender nonconformity will be discussed in more depth at the end of this section.

Interpersonal and Institutional Discrimination

Although anti-lesbian/gay discrimination is not only produced by status generalization (e.g., another source is simply hatred or disgust), it is important to briefly discuss this discrimination since it coexists in societies where the states of sexual orientation are differentially valued. Those who possess the low state of any characteristic are often targets of blatant and/or subtle discrimination.

Given prevalence of negative stereotypes of and attitudes toward gays and lesbians, it is not surprising that a significant number of gays and lesbians have been and continue to be targets of physical assault, such as having objects thrown at them or being punched, hit, kicked, or beaten. They also suffer verbal abuse such as being ridiculed or called insulting names (Herek 1989; Comstock 1991; Berrill 1992). Comstock (1991) concluded that slightly over half of socially active lesbians and gay men (i.e., those who frequent organizations, such as political and social gay and lesbian organizations, bookstores, or bars, where questionnaires are typically distributed) experience some type of physical violence or potential threat of violence based on their sexual orientation. Gay men are more likely to be victims of physical violence than lesbians (Berrill 1992). Lesbians of color are more likely to be victimized, particularly chased or followed, than white lesbians (Comstock 1991, p. 46).

In addition, the belief that heterosexuals are "better" and more worthy than homosexuals is evidenced by the preponderance of anti-lesbian/gay discrimination within most, if not all, of our institutions (e.g., economic, political, military, religious, and family). Sexual orientation is nearly always excluded from the list of protected minorities in civil rights legislation (Herek 1991, 1993; Rivera 1991; Garnets and Kimmel 1993). Except in two states (Wisconsin and Massachusetts) and a few dozen cities, legal protections do not exist for gays and lesbians in employment, housing, or services (Rivera 1991; Blumenfeld 1992; Herek 1993). Also, for the majority of the gay population, being "out" on the job is extremely risky; loss of job, denial of promotion, and awkward daily interactions with coworkers are possible consequences (Rubin 1984; Woods 1993).

Further, gay and lesbian relationships are clearly not given the same legitimacy as heterosexual relationships, as evidenced by the fact that these relationships have no legal or family status (Rivera 1991; Herek 1993). Gays and lesbians are not allowed to legally marry currently in any state and, therefore, do not benefit from the numerous economic advantages that straight married couples take for granted and enjoy (Rubin 1984; Herek 1991). For example, in most private and public organizations, except those with domestic partnership policies, gay employees cannot obtain employment health benefits or recreational privileges for their partners. Further, gay and lesbian partners can be barred from visiting a sick or dying lover in the hospital since they are not considered an immediate family member (Goodman, Lakey, Lashof, and Thorne 1983; Herek 1991; Blumenfeld 1992). Lesbian mothers have difficulty keeping custody of their children after divorce because it is often assumed that they are emotionally unstable or unable to assume the maternal role (Rivera 1991; Falk 1993). And adoption by gay men and lesbians is still not easily accepted. The level of discrimination by the courts in child custody cases and adoption cases varies by state.

In addition, one obvious American institution where antigay discrimination is clearly stated and institutionally supported is the military (Rivera 1991). Thousands of people have been removed from the military service for "being against regulation," (i.e., for saying that they are lesbian or gay), no matter how exemplary their service records (Herek 1991; Humphrey 1990; Shilts 1993). Currently, under the Clinton Administration, a new policy was implemented called "Don't Ask, Don't Tell, Don't Pursue." This policy prevents recruitment officers from directly asking if someone is homosexual. Of course, once in the military, gays and lesbians must keep their sexual orientation a secret; that is, they must keep one of their most important identities to themselves and continue to live in the closet.

Clearly gays and lesbians face actual and/or the threat of institutional and interpersonal discrimination on a daily basis simply because of their sexual orientation. Given prevalence of this discrimination, in addition to the negative stereotypes of and attitudes toward gay men and lesbians based largely on violation of gender roles, it is clear that there is a more favorable overall evaluation of heterosexuals than gays and lesbians in our society.

Sexual Orientation and General Expectation States

In order for sexual orientation to be a diffuse status characteristic, states of the characteristic must not only be differentially valued in society, but must also be linked to general performance expectations. Only one study thus far has examined the relationship between sexual orientation and general expectations for competence.

Based on Webster and Driskell's (1983) previous work on beauty as a status characteristic, Hysom (1994) developed descriptions of individuals and a questionnaire to test the effects of sexual orientation and occupation on general expectation states. In the descriptions, individuals were said to be in the military. Information on sexual orientation (a heterosexual male and a homosexual male) and occupation (computer systems analyst and dishwasher) for the individuals systematically varied across conditions. Other information on age, educational level, and physical descriptions was held constant across conditions. Two separate measures of general performance expectations were used to create two dependent variables. The first set of questions indirectly measured general expectations by asking subjects which individual they would rather have as a leader of their stational unit and of their combat unit. The second set of questions, adapted from Webster and Driskell (1983), directly measured general expectation states by asking respondents to compare individuals on a number of general skills (e.g., intelligence, high school GPA, and military and general knowledge).

Results on both measures supported the contention that sexual orientation is related to general expectation states in the predicted direction (at least in the case of straight and gay men). Respondents formed higher general expectations for men with a heterosexual orientation than for men with a homosexual orientation.

As suggested by Hysom (1994), further research is obviously needed to strengthen the claim that sexual orientation often functions as a diffuse status characteristic. In particular, the relationship between sexual orientation and general expectation states should be examined for women. This initial study, however, provides a critical starting point for research on sexual orientation and status. (See also a new study by Webster and Hysom 1995.)

Sexual Orientation and Gender Nonconformity

Although there is no inherent connection between sexual orientation and gender nonconformity, popular stereotypes associate lesbianism with masculinity and male homosexuality with effeminacy (Stokes, Kilmann, and Wanlass 1983; Herek 1984, 1991; Weston 1993; Woods 1993). Gay men are presumed to manifest characteristics that are culturally defined as "feminine"— for example, passive, sensitive, dependent, emotional, and artistic; lesbians are perceived to manifest "masculine" characteristics—for example, aggressive, competitive, assertive, independent, and mechanical (Broverman, Vogel, Broverman, Clark, and Rosenkrantz 1972; Baird 1976; Herek 1984, 1991; Kite and Deaux 1986, 1987). Specifically, traits typically associated with gay men include theatrical, unaggressive, gentle, weak, unmasculine, emotional, passive, not a good leader, sensitive, and less dominant and competitive than straight men (Gross, Green, Storch, and Vanyur 1980; Taylor 1983; Herek 1991). Some stereotypical nonverbal behavior includes swish gait, limp wrist, immaculate hair style, lisp, a high level of gesturing, and a feminine demeanor and appearance (Plummer 1975; Berger, Hank, Rauzi, and Simkins 1987). Lesbians are perceived as more dominant, direct, forceful, strong, and nonconforming than straight women. The stereotype also includes manlike demeanor such as walking like a man, "butch" appearance such as short hair cut and no makeup, and a very assertive attitude (Berger, Hank, Rauzi, and Simkins 1987; Herek 1991; Pelligrini 1992).[2]

In order to test whether people rate lesbians as more masculine/less feminine than straight women and gay men as more feminine/less masculine than straight men, Taylor (1983) examined homosexual stereotypes "within the domain of masculine and feminine traits" (p. 43). One hundred and three adults (ages 17 to 64) were provided with four copies of the Personality Attributes Questionnaire (Spence and Helmreich 1978) to assess masculine and feminine stereotypes. Each copy bore one of the four group labels, "men" (heterosexual men), "women" (heterosexual women), "male homosexuals," or "lesbians." Respondents rated each group according to traits they thought were associated with each category. Results showed that people attach more masculine traits to lesbians than to straight women and more feminine traits to gay men than to straight men.[3]

Gay men and lesbians, then, are perceived as violating gender roles that threatens the traditional gender role structure. It is not surprising, therefore, that Schwanberg (1993), who examined 45 studies of attitudes toward gay men and lesbians, concluded that attitudes are, in general, quite negative. In addition, disapproval of gender role deviation is evident in studies of parents' and peers' reactions to children who manifest opposite gender personality characteristics. When children behave in ways that run counter to traditional roles, their activities are often discouraged by parents (Martin 1990). There is considerable pressure to change their behavior by parents and peers (Green 1987). Boys who routinely engage in "feminine" activities, however, are viewed more negatively than girls who engage in more "masculine" activities (Martin 1990; Garnets and Kimmel 1993). "Sissies" are more negatively evaluated than "tomboys" for several reasons. First, culturally defined characteristics of masculinity are more highly valued than culturally defined feminine characteristics. Second, girls are believed to more easily grow out of the gender nonconforming behavior, whereas boys may not. Since there exists the perception that this nonconforming behavior is associated with becoming "homosexual," there is a fear that boys will grow up to be "homosexual" (Block 1983; Martin 1990). As Garnets and Kimmel (1993) note, the stigma of homosexuality is a very effective way of reinforcing gender roles.

Similar to the results on perceptions of child deviation from gender norms, Page and Yee (1985) found that the stereotypical traits associated with gay men were rated more negatively than the stereotypical traits of lesbians. Again, this makes sense given that femininity is less highly valued than masculinity.

In addition, there is some evidence that, among heterosexuals, men have more negative attitudes toward "homosexuals" and less tolerance of gender nonconformity than women, although the evidence is inconsistent and may be generalizable only to white samples (Kite 1984; Herek 1988; Whitley 1988; Garnets and Kimmel 1993; Herek and Glunt 1993; Schwanberg 1993; Eliason 1994). For example, Martin (1990) found that women were more accepting of gender nonconformity in children than men were. Lieblich and Friedman (1985), in their study of 170 American and Israeli students' attitudes toward gay men and lesbians, found that the American and Israeli male students were significantly more prejudiced toward both gay men and lesbians than the female students. Herek and Glunt (1993), in a national AIDS telephone survey, found that men were significantly more likely than women to express negative attitudes toward gay men. The literature on the relationship between homosexuality and perceived gender nonconformity is unclear, but I use these ideas to develop a theoretical argument about the activation of sexual orientation and its consequences in small groups.

ACTIVATION OF SEXUAL ORIENTATION AND GENDER

Given that sexual orientation is a diffuse status characteristic, how does it operate in small group settings? How does it become salient and, once activated, how does it affect interaction and the development of the status hierarchy? I begin this analysis by first considering the issue of visibility and its implications for activation of sexual orientation. I then discuss how sexual orientation may be activated and how the status characteristic, gender, may also become salient under certain conditions.

Status characteristics such as race and age are usually readily distinguishable by the presence of obvious categorical cues, such as skin color and other visible indicators of race or age, or less obvious cues such as manner of dress. Sexual orientation is more difficult to assess in many situations for several reasons. First, many gay men and lesbians themselves go to great lengths to preserve the secret of their sexual orientation because often the costs or potential costs of disclosure, including fear of discrimination, are high.[4] Some gays and lesbians, for example, learn to manage their identities at work by paying close attention to nuances of appearance and gesture and to the disclosure of information about their personal lives (Woods 1993). In addition, Weston (1993) argues that individuals who are attempting to pass for straight must be particularly careful to manage the symbols that are more specifically linked to gender than sexuality per se since many people confuse homosexuality with gender nonconformity. Second, lesbians and gay men live in a world where a "heterosexual assumption" is made in daily interactions. This is an implicit assumption that everyone we meet is heterosexual unless we are told otherwise (Ponse 1978; Goodman, Lakey, Lashof, and Thorne 1983; Blumenfeld 1992; Woods 1993).

Although sexual orientation is often hidden in many contexts, there are several ways in which it may be activated in a group setting. First, categorical cues, both indicative and expressive, may be used to identify sexual orientation (Ridgeway, Berger, and Smith 1985; Berger, Webster, Ridgeway, and Rosenholtz 1986). Indicative categorical cues may be gay identifying symbols worn on the person such as a pink or black triangle pin, or t-shirts that reference gay or lesbian identity (e.g., a t-shirt with the saying "I'm not a lesbian, but my girlfriend is," or "Silence = Death"). Other indicative cues include making statements that disclose the identity such as, "As a gay man, I....," or providing knowledge that indicates membership in gay/lesbian groups. In these cases, the individual is at least not unwilling to be publicly identified as gay.

Expressive categorical cues that may activate sexual orientation are connected to the stereotypes of gays and lesbians and their relation to gender stereotypes.[5] These include actions that are presumably associated with homosexuality, such as certain speech style, mannerisms, and appearance (e.g., hair style, dress). As discussed earlier, gay men are stereotyped as effeminate and lesbians as masculine in behavior and appearance. People may use these stereotypical beliefs to infer gay or lesbian identity (Plummer 1975; Weston 1994). In many cases, however, observers may misidentify an individual's sexual orientation, either by incorrectly perceiving someone as lesbian or gay when she or he is not or not perceiving someone as lesbian or gay when she or he actually is.

An interesting study by Berger, Hank, Rauzi, and Simkins (1987) showed that the popular stereotypes of gay males being effeminate and lesbians being "butch" did not aid 143 gay and straight participants in their identification of 24 gay and straight target individuals. The subjects were asked to watch the target individuals engage in semi-structured two to three minute interviews, and then determine if each target was gay or straight. They were also asked to briefly give a reason for the determinations they made of the targets' sexual orientation.

Results showed that participants correctly identified individuals of their own sexual orientation no better than they did any other group. Straight women were better than gay men at identifying straight men. Lesbians were better than straight women at identifying gay men, but even they had difficulty identifying other lesbians. As important, criteria used to identify individuals' sexual orientation show that gay and lesbian stereotypes were used to determine identification. Some criteria used were masculine or feminine appearance (very butch or feminine demeanor), attitude (e.g., confident manner, ease with self, hostile demeanor, assertiveness, low self-esteem), general appearance (e.g., hairstyle and dress), gestures and posture (e.g., way of sitting and walking and hand movements), and speech and voice (e.g., way of talking, choice of words, pattern of speech, and lisp). An important implication of this study is that people will often make determinations about an individual's sexual orientation based on stereotypes they hold about gender and sexual orientation, even though they may be incorrect in their assessment.

Categorical cues, both indicative and expressive, seem to be used to make attributions about a person's sexual orientation. In task group settings, one can see how both indicative and expressive categorical cues might be used to activate sexual orientation. I suggest in addition, however, that gender may become salient in situations where expressive categorical cues are displayed in addition to indicative categorical cues. For example, consider a situation where two males are assigned to work together on a task. In this case let us say that they are both aware of each other's sexual orientation through indicative categorical cues (where one is gay and the other is straight) and they do not differ on any other status characteristic. Let us further assume that the gay member also displays overt stereotypical cues (i.e., expressive categorical cues), such as lispy manner of speech and effeminate mannerisms and body movements). Sexual orientation should be very salient in this case.

Further, however, it is reasonable to assume that gender will also be activated in this situation even though this is a same-sex dyad. Why might this be so? Given that adherence to traditional gender roles is highly valued in our society and that a common stereotype of gays is linked to gender role deviation, when the straight male perceives the gay male as being more feminine than masculine in his expressive characteristics, gender roles will become salient in the situation. Not only is the gay male gay, he also violates gender roles by displaying "feminine" cues.

Gender, then, should become salient when expressive categorical cues support gay and lesbian stereotypes. For example, in same-sex groups where gay males "give off feminine" expressive cues, sexual orientation and gender will be activated; in same-sex groups where lesbians "give off masculine" expressive cues, again both characteristics should be activated. When sexual orientation is activated only through indicative categorical cues, however, where expressive cues are not consistent with stereotypes, sexual orientation should be salient but gender should not be activated.

In order to flesh out the implications of sexual orientation in problem-solving groups, I will provide specific predictions for situations that vary in how sexual orientation is activated. The type of cues that activate sexual orientation is the key to whether gender will also be activated and ultimately how sexual orientation affects interaction.

PREDICTIONS IN SAME-SEX DYADS

In accord with Berger, Wagner, and Zelditch's (1989) approach to theoretical development and strategies, I begin this preliminary analysis of the effects of sexual orientation by examining the simplest unit of analysis, the same-sex dyad. I offer a set of predictions for three possible types of situations involving same-sex dyads where members differentiate only on sexual orientation. These situations vary only in how sexual orientation is activated. In addition, the task is gender neutral in all three situations. This is important since, as we shall see, gender is salient in two of the three situations.

Situation One

In situation one, sexual orientation is activated through indicative categorical cues only. Expressive cues are neutral. Both members are aware of each other's sexual orientation, but the gay member is not "giving off" stereotypical cues. In this case, sexual orientation should become salient in both male and female dyads. Gender, however, will not be activated since there are no nonverbal cues that link the individual to gender stereotypes.

Male Dyads

In male dyads, then, the gay male should have an expectation disadvantage compared to the straight male since he possesses the low state of sexual orientation. He should form lower performance expectations for himself compared to the straight member; the straight member should have higher performance expectations for himself compared to the gay member. As a result, we would expect that the straight male will contribute more ideas and have slightly more influence in decisions than the gay male.

Female Dyads

Similarly, in female dyads, the lesbian member should have an expectation disadvantage compared to the straight woman. It may be, however, that if women are more tolerant or accepting of "homosexuals," sexual orientation, although activated, may not be as salient in these groups as in the male dyads. Status generalization effects of sexual orientation may be weaker for women than for men. If this is the case, these groups should look very similar to typical female peer dyads. If not, the lesbian should form lower performance expectations for herself compared to the straight member; the straight member should have higher performance expectations for herself compared to the lesbian member.

Situation Two

In the second situation, sexual orientation is indicated through both indicative and expressive categorical cues and expressive cues are consistent with stereotypes. Both members are aware of each other's sexual orientation and the gay member is "giving off" expressive cues that are consistent with the gay stereotype (i.e., "feminine" cues for men and "masculine" cues for women). In this case, then, I propose that both sexual orientation and gender should become salient. If this does occur, the implications for interaction are considerably different for all-male and all-female dyads.

Male Dyads

In male dyads, the gay male will be doubly disadvantaged. He will possess the lower state of sexual orientation and he will possess the lower state of gender, since "feminine cues" are less highly valued than "masculine" cues and, when displayed by men, are even more devalued than when displayed by women. In this particular dyad, both sexual orientation and gender will be activated and the gay member will possess low states of both characteristics. Gay men in this second situation, then, should have a significantly larger disadvantage than gay men in situation one because they possess the low states of two characteristics rather than one.

Female Dyads

Similar to male dyads, sexual orientation and gender will be activated, but the implications are quite different. The lesbian in this case will possess the lower state of sexual orientation, but in contrast to the gay man, she will possess the higher state of gender. That is, since her expressive cues are stereotypically "masculine" and since masculine characteristics are highly valued, she will possess the higher state of gender. In this case, then, each member possesses one high and one low state of the characteristics. This suggests that the members should be status equals in the situation. It is possible, however, that the lesbian has an expectation advantage if sexual orientation is a weaker characteristic for women than for men.

Situation Three

In a third situation, sexual orientation is indicated only through expressive categorical cues which are consistent with stereotypes. There is no direct claim to possessing a particular state of sexual orientation, but there are expressive cues that indirectly indicate sexual orientation. In this case, it is possible that the display of expressive cues may make salient both sexual orientation and gender. If this is so, then the predictions should be similar to those for situation two. In male dyads, gay men will possess the low states of both characteristics, giving them a distinct expectation disadvantage compared to straight men. In female dyads, lesbians will possess the low state of sexual orientation, but the high state of gender and, therefore, should be status equals with their straight partners.

In summary, the predictions from these situations, when examined together, are the following:

1. In male dyads, gay men will have a significantly greater expectation disadvantage in situations where their display of expressive categorical cues are stereotypical than in situations where only indicative categorical cues are displayed.
2. In female dyads, lesbians will have either equal status or will have a significantly greater expectation advantage in situations where their display of expressive categorical cues are stereotypical than in situations where only indicative categorical cues are displayed.

A test of these predictions will not only help us understand how sexual orientation and gender are related to each other and operate in groups, but will also allow us to test the relative importance of indicative categorical and expressive cues. Very little research has been done on testing the strength of these types of cues on performance expectations, and ultimately, behavior.

How to Test the New Ideas[6]

In the theoretical formulation stated earlier, I have argued that sexual orientation may be identified through indicative categorical cues, stereotypical expressive categorical cues, or both and that gender may become salient when expressive cues are exhibited. Two interrelated problems with this formulation must be addressed, however, when considering the empirical testing of these predictions. The first issue concerns the possibility of confounding the effects of expressive categorical with expressive task cues on performance expectations. Expressive categorical cues provide information about the diffuse status characteristic a member possesses; expressive task cues provide information on the ability of the member. To address this potential problem, initial tests of the activation of sexual orientation must be sure to manipulate only expressive categorical cues, while keeping task cues neutral, since they provide different types of information.

A second related issue concerns how to determine whether or not gender, in addition to sexual orientation, really does become salient when stereotypic expressive cues are exhibited. It may be the case, for example, that when expressive categorical cues indicating sexual orientation are present simultaneously with indicative categorical cues, the salience of sexual orientation is intensified rather than gender becoming salient.

To assess the idea that gender may become salient when stereotypical expressive categorical cues are displayed in addition to indicative categorical cues, I outline an experimental study that compares performance expectations for self and other in the following set of conditions with same-sex dyads. In the first condition, similar to Situation One, orientation is identified only through indicative categorical cues. In this case, imagine both members know each other's sexual orientation through some type of indicative categorical cue, where one is straight and the other is gay or lesbian. In the second condition, expressive categorical cues associated with gay and lesbian stereotypes would be manipulated, in addition to indicative categorical cues. In male dyads, the gay member displays "feminine" cues and in female dyads the lesbian displays "masculine" cues through, for example, dress, body movements, physical appearance, and way of sitting. Expressive task cues (e.g., speech fluency and hesitancy, tone, and duration of eye contact) would remain constant across conditions. We then should be able to determine the effects of indicative and expressive categorical cues on performance expectations.

In male dyads, if performance expectations for the gay man are lower in condition two than in condition one, this suggests that expressive categorical cues do have an effect above and beyond indicative categorical cues. We will not be able to determine, however, the cause of the change in performance expectations; that is, is it simply due to an intensification of sexual orientation or has gender become activated as well.

In female dyads, however, we should be able to determine whether gender becomes salient in addition to sexual orientation or whether sexual orientation is simply intensified. If performance expectations for the lesbian are lower in condition two than in condition one, this suggests that sexual orientation is probably simply intensified by the addition of expressive categorical cues. If, on the other hand, performance expectations for her are equal or higher in condition two than in condition one, we have evidence that gender is activated and the lesbian, because of "masculine cues" has the higher state. Comparing conditions that vary indicative and expressive categorical cues allows us to differentiate between the idea that sexual orientation is intensified when expressive categorical cues are exhibited or, instead, two characteristics, sexual orientation and gender, become salient.

Once the effects of expressive categorical cues are clarified, then the effects of the combination of expressive categorical and task cues associated with gay and lesbian stereotypes should be examined. Characteristics associated with gay and lesbian stereotypes (and linked to gender stereotypes) involve both expressive categorical cues as described earlier *and* expressive task cues (Berger, Hank, Rauzi, and Simkins 1987). For example, characteristics associated with demeanor are also part of the gay and lesbian stereotypes. Lesbians are expected to be assertive, directive, aggressive, and have body language that represents a no nonsense attitude. Gay men are expected to have a more "feminine" demeanor, including being more passive, unaggressive, and less competitive. Therefore, a combination of expressive categorical and task cues may activate sexual orientation in task groups. Although more complicated, the role of expressive task cues in the activation of sexual orientation will need to be explored.

CONCLUSION

Sexual orientation has, thus far, been left out of any discussion of status characteristics theory. In this paper, I have argued that it is clearly a diffuse status characteristic in our society, evidenced by the existence of negative stereotypes of and attitudes toward gays and lesbians based largely, though not completely, on perceived gender role deviation, and some preliminary evidence that shows a link between the states of sexual orientation and distinct sets of general expectation states.

I have examined the role that indicative categorical and expressive categorical cues play in the activation of sexual orientation in task groups. The most interesting consequence of the activation of sexual orientation through expressive cues is the possibility of the simultaneous activation of a second characteristic, gender, in same-sex dyads. Based on this possibility, I provided some preliminary predictions on the effects of sexual orientation in same-sex dyads. I then discussed two critical issues involved in testing these predictions, particularly the importance of separating the effects of expressive categorical from expressive task cues on the forming of performance expectations, and suggested an experimental study to test these predictions.

An examination of how sexual orientation affects small group interaction should tell us more about how indicative and expressive status cues operate in groups and how diffuse status characteristics are activated. It should also advance our assessment of the exact role, if any, gender plays in groups where members differentiate on sexual orientation. Finally, by understanding the way in which sexual orientation operates as a diffuse status characteristic in interaction we can begin to examine how interaction techniques developed for other status characteristics (e.g., Cohen, Lotan, and Catanzarite 1988; Cohen 1993) may be adapted to overcome undesirable features of status generalization based on sexual orientation.

ACKNOWLEDGMENT

I am indebted to Eve Hayworth and Murray Webster, Jr. for their insightful comments on earlier versions of this paper. I also thank Rebecca Ford and the editors of this volume for their helpful suggestions.

NOTES

1. The term sexual orientation is used because it best encompasses various states of this characteristic: that is, homosexuality (gay/lesbian), heterosexuality (straight), and bisexuality.

2. Of course, gay and lesbian stereotypes are not limited to gender nonconformity, but this does represent a popular stereotype for many people. In addition, obviously not all heterosexuals hold these gay and lesbian stereotypes, just as not all men hold negative stereotypes of women. There is sufficient evidence to suggest, however, that popular stereotypes of gays and lesbians are linked to gender nonconformity.

3. Interestingly, Finlay and Scheltema (1991) examined self-reports of lesbian and straight women and gay and straight men on measures of masculinity, femininity, and androgyny, using the Personality Attributes Questionnaire. They found that lesbians scored higher on the masculine scale than straight women but similar on the feminine and androgynous scales. Gay men scored lower on the masculine scale than straight men but similar on the feminine and androgynous scales. Specifically, lesbians rated themselves higher on independence, self-confidence, and ability to make decisions than straight women. Gay men rated themselves lower on competitiveness, self-confidence, and ability to make decisions than straight men.

4. Examples of costs are discrimination in employment such as the denial of the right to practice certain professions or the denial of promotions in the workplace, denial of child custody, being a potential or actual target of antigay/lesbian hate crimes, and fear of rejection by and alienation from family and friends (Goodman et al. 1983; Herek 1989; Hall 1989; Wells and Kline 1987; Woods 1993).

5. Of course, common perceptions of the relationship between gender and sexuality are varied and complex, but it is beyond the scope of this paper to address this complexity. At this stage, I am only interested in how sexual orientation, in general, may be activated through categorical and expressive cues and what this may mean for task group interaction.

6. I owe special thanks to my colleague, Murray Webster, Jr. for his invaluable help in providing the foundation for this portion of the paper.

REFERENCES

Baird, J. E. 1976. "Sex Differences in Group Communication: A Review of Relevant Research." *Quarterly Journal of Speech* 62: 179-92.

Berger, G., L. Hank, T. Rauzi, and L. Simkins. 1987. "Detection of Sexual Orientation by Heterosexuals and Homosexuals." *Journal of Homosexuality* 13: 83-100.

Berger, J., T. L. Conner, and M. H. Fisek. 1974. *Expectation States Theory: A Theoretical Research Program*. Cambridge, MA: Winthrop.

Berger, J., M. H. Fisek, R. Z. Norman, and M. Zelditch, Jr. 1977. *Status Characteristics in Social Interaction: An Expectation States Approach*. New York: Elsevier.

Berger, J., S. Rosenholtz, and M. Zelditch, Jr. 1980. "Status Organizing Processes." *Annual Review of Sociology* 6: 479-508.

Berger, J., M. Webster, Jr., C. Ridgeway, and S. Rosenholtz. 1986. "Status Cues, Expectations, and Behavior." Pp. 1-22 in *Advances in Group Processes* (Volume 3), edited by E. Lawler. Greenwich, CT: JAI Press.

Berger, J., D. Wagner, and M. Zelditch, Jr. 1989. "Theory Growth, Social Processes and Metatheory." Pp. 19-42 in *Theory Building in Sociology: Assessing Theoretical Cumulation*, edited by J. Turner. Newbury Park, CA: Sage.

Berrill, K. T. 1992. "Anti-gay Violence and Victimization in the United States: An Overview." Pp. 19-64 in *Hate Crimes: Confronting Violence Against Lesbians and Gay Men*, edited by G. Herek and K. Berrill. Newbury Park, CA: Sage.

Block, J. H. 1983. "Differential Premises Arising from Differential Socialization of the Sexes: Some Conjectures." *Child Development* 54: 1335-54.

Blumenfeld, W. J., (ed.). 1992. *Homophobia: How We All Pay the Price*. Boston: Beacon.

Broverman I. K., S. R. Vogel, D. M. Broverman, F. E. Clark, and P. S. Rosendrantz. 1972. "Sex Role Stereotypes: A Current Appraisal." *Journal of Social Issues* 28: 59-78.

Cohen, E. G. 1993. "From Theory to Practice: The Development of an Applied Research Program." Pp. 385-415 in *Theoretical Research Programs: Studies in the Growth of Theory*, edited by J. Berger and M. Zelditch. Stanford, CA: Stanford University Press.

Cohen, E. G., R. Lotan, and L. Catanzarite. 1988. "Can Expectations for Competence be Altered in the Classroom?" Pp. 27-54 in *Status Generalization: New Theory and Research*, edited by M. Webster and M. Foschi. Stanford, CA: Stanford University Press.

Comstock, G. D. 1991. *Violence Against Lesbians and Gay Men*. New York: Columbia University Press.

Deaux, K. and L. L. Lewis. 1984. "Structure of Gender Stereotypes: Interrelationships Among Components and Gender Label." *Journal of Personality and Social Psychology* 46: 991-1004.

Eliason, M. J. 1994. "Attitudes About Lesbians and Gay Men: A Review and Implications for Social Service Training." *Journal of Gay and Lesbian Social Services*. In press.

Elze, D. 1992. "It Has Nothing to do With Me." Pp. 95-113 in *Homophobia: How We All Pay the Price*, edited by W. Blumenfeld. Boston: Beacon.

Falk, P. J. 1993. "Lesbian Mothers: Psychological Assumptions in Family Law." Pp. 420-436 in *Psychological Perspectives on Lesbian and Gay Male Experiences*, edited by L. D. Garnets and D. C. Kimmel. New York: Columbia University Press.

Finlay, B. and K. Scheltema. 1991. "The Relation of Gender and Sexual Orientation to Measures of Masculinity, Femininity, and Androgyny: A Further Analysis." *Journal of Homosexuality* 21: 71-85.

Garnets, L. D. and D. C. Kimmel. 1993. "Lesbian and Gay Male Dimensions in the Psychological Study of Human Diversity." Pp. 1-52 in *Psychological Perspectives on Lesbian and Gay Male Experiences*, edited by L. D. Garnets and D. C. Kimmel. New York: Columbia University Press.

Goffman, I. 1959. *The Presentation of Self in Everyday Life*. Garden City, NY: Doubleday/ Anchor.

Goodman, G., G. Lakey, J. Lashof, and E. Thorne. 1983. *No Turning Back: Lesbian and Gay Liberation for the '80s*. Philadelphia: New Society Publishers.

Green, R. 1987. *The "Sissy Boy Syndrome" and the Development of Homosexuality*. New Haven, CT: Yale University Press.

Gross, A, S. Green, J. Stork, and J. Vanyur. 1980. "Disclosure of Sexual Orientation and Impressions of Male and Female Homosexuals." *Personality and Social Psychology Bulletin* 6: 307-14.

Hayworth, E. L. 1991. "Sexual Orientation as a Diffuse Status Characteristic." Unpublished paper.

Herdt, G. 1989. "Gay and Lesbian Youth, Emergent Identities, and Cultural Scenes at Home and Abroad." *Journal of Homosexuality* 17: 1-42.

Herek, G. M. 1984. "Beyond 'Homophobia': A Social Psychological Perspective on Attitudes Toward Lesbians and Gay Men." *Journal of Homosexuality* 10: 1-21.

_____. 1988. "Heterosexuals' Attitudes Toward Lesbians and Gay Men: Correlates and Gender Differences." *The Journal of Sex Research* 25: 451-77.

——————. 1989. "Hate Crimes Against Lesbians and Gay Men: Issues for Research and Policy."
    American Psychologist 44: 948-955.
——————. 1991. "Stigma, Prejudice, and Violence Against Lesbians and Gay Men." Pp. 60-80
    in *Homosexuality: Research Implications for Public Policy*, edited by J. Gonsiorek and
    J. Weinrich. Newbury Park, CA: Sage.
——————. 1993. "The Context of Antigay Violence: Notes on Cultural and Psychological
    Heterosexism." Pp. 89-107 in *Psychological Perspectives on Lesbian and Gay Male
    Experiences*, edited by L. D. Garnets and D. C. Kimmel. New York: Columbia University
    Press.
Herek, G. M. and E. K. Glunt. 1993. "Interpersonal Contact and Heterosexuals' Attitudes Toward
    Gay Men: Results from a National Survey." *The Journal of Sex Research* 30: 239-44.
Humphrey, M. 1990. *My Country, My Right to Serve*. New York: Harper Collins.
Hysom, S. J. 1994. "Sexual Orientation as a Diffuse Status Characteristic." Unpublished
    manuscript. Available from department of Sociology, UNCC.
Kite, M. E. 1984. "Sex Differences in Attitudes Toward Homosexuals: A Meta-Analytic Review."
    Journal of Homosexuality 10: 69-81.
Kite, M. E. and K. Deaux. 1986. "Attitudes Toward Homosexuality: Assessment and Behavioral
    Consequences." *Basic and Applied Social Psychology* 7: 137-62.
——————. 1987. "Gender Belief Systems: Homosexuality and the Implicit Inversion Theory."
    Psychology of Women Quarterly 11: 83-96.
Lieblich, A. and G. Friedman. 1985. "Attitudes Toward Male and Female Homosexuality and
    Sex-role Stereotypes in Israeli and American Students." *Sex Roles* 12: 561-70.
Lockheed, M. E. and K. P. Hall. 1976. "Conceptualizing Sex as a Status Characteristic:
    Applications to Leadership Training Strategies." *Journal of Social Issues* 32: 11-24.
Martin, C. L. 1990. "Attitudes and Expectations about Children with Nontraditional and
    Traditional Gender Roles." *Sex Roles* 22: 151-65.
Morgan, K. S. and L. S. Brown. 1993. "Lesbian Career Development, Work Behavior, and
    Vocational Counseling." Pp. 267-296 in *Psychological Perspectives on Lesbian and Gay
    Male Experiences*, edited by L. D.Garnets and D. C. Kimmel. New York: Columbia
    University Press.
Page, S. and M. Yee. 1985. "Conception of Male and Female Homosexual Stereotypes Among
    University Undergraduates." *Journal of Homosexuality* 12: 109-118.
Pellegrini, A. 1992. "S(h)ifting the Terms of Hetero/Sexism: Gender, Power, Homophobias." Pp.
    39-56 in *Homophobia: How We All Pay the Price*, edited by W. Blumenfeld. Boston:
    Beacon.
Piliavin, J. A. and R. R. Martin. 1978. "The Effects of the Sex Compositions of Groups on Style
    of Social Interaction." *Sex Roles* 4: 281-96.
Plummer, K. 1975. *Sexual Stigma: An Interactionist Account*. London: Routledge and Kegan
    Paul.
Ponse, B. 1978. *Identities in the Lesbian World: the Social Construction of the Self.* Westport,
    CT: Greenwood.
Preston, K. and K. Stanley. 1987. "What's the Worst Thing..?" Gender-directed Insults." *Sex Roles*
    17: 209-19.
Pugh, M. D. and R. Wahrman. 1985. "Inequality of Influence in Mixed-sex Groups." Pp. 142-
    52 in *Status, Rewards, and Influence: How Expectations Organize Behavior*, edited by
    J. Berger and M. Zelditch. San Francisco: Jossey-Bass.
Ridgeway, C. 1991. "The Social Construction of Status Value: Gender and Other Nominal
    Characteristics." *Social Forces* 70: 367-86.

Ridgeway, C., J. Berger, and L. Smith. 1985. "Nonverbal Cues and Status: An Expectation States Approach." *American Journal of Sociology* 90: 955-978.

Rivera, R. R. 1991. "Sexual Orientation and the Law." Pp. 81-100 in *Homosexuality: Research Implications for Public Policy*, edited by J. Gonsiorek and F. Weinrich. Newbury Park, CA: Sage.

Rubin, G. 1984. "Thinking Sex." Pp. 267-319 in *Pleasure and Danger: Exploring Female Sexuality*, edited by Carole Vance. London: Routledge and Kegan Paul.

Schwanberg, S. L. 1993. "Attitudes Toward Gay Men and Lesbian Women: Instrumentation Issues." *Journal of Homosexuality*: 99-136.

Shilts, R. 1993. *Conduct Unbecoming: Lesbians and Gays in the U.S. Military: Vietnam to the Persian Gulf.* New York: St. Martin's Press.

Spence, J. T. and R. L. Helmreich. 1978. *Masculinity and Femininity.* Austin: University of Texas Press.

Stokes, K., P. R. Kilmann, and R. Wanlass. 1983. "Sexual Orientation and Sex Role Conformity." *Archives of Sexual Behavior* 12: 427-433.

Strodtbeck, F. L. and R. D. Mann. 1956. "Sex Role Differentiation in Jury Deliberation." *Sociometry* 19: 3-11.

Taylor, A. 1983. "Conceptions of Masculinity and Femininity as a Basis for Stereotypes of Male and Female Homosexuals." *Journal of Homosexuality* 3: 37-53.

Webster, M., Jr. and J. E. Driskell. 1983. "Beauty as Status." *American Journal of Sociology* 89: 140-65.

Webster, M., Jr. and S. J. Hysom. 1995. "Sexual Orientation as Status." Submitted to Group Processes Section, ASA.

Wells, J. W. and W. B. Kline. 1987. "Self-disclosure of Homosexual Orientation." *Journal of Social Psychology* 127: 191-7.

Weston, K. 1993. "Do Clothes Make the Woman?: Gender, Performance Theory, and Lesbian Eroticism." *Genders* 17: 1-21.

Whitley, B. E. Jr. 1988. "Sex Differences in Heterosexuals Attitudes Toward Homosexuals: It Depends Upon What You Ask." *Journal of Sex Research* 24: 287-91.

Wood, W. and S. Karten. 1986. "Sex Differences in Interactional Style as a Product of Perceived Sex Differences in Competence." *Journal of Personality and Social Psychology* 50: 341-347.

Woods, J. D. 1993. *The Corporate Closet: The Professional Lives of Gay Men in America.* New York: The Free Press.

THE DETERMINANTS OF TOP
MANAGEMENT TEAMS

Karen A. Bantel and Sydney Finkelstein

ABSTRACT

Conceptualizing the top management team as the strategic decision-making unit for the firm, the team's key attributes are described in terms of demographic mean characteristics, demographic diversity, and size. A model of the key antecedents of these attributes is developed, focusing on environmental, strategy, and organizational determinants.

Top management teams have become the subject of intense interest in recent years among both business academics and managers. Faced with increasingly competitive and rapidly changing industry conditions, top executives must confront myriad pressures from a variety of sources. They also must provide creative leadership, careful to ensure that their strategic perspectives and actions are appropriate to today's changing environment, while at the same time being held highly accountable to a range of stakeholders, often with

Advances in Group Processes, Volume 12, pages 139-165.
Copyright © 1995 by JAI Press Inc.
All rights of reproduction in any form reserved.
ISBN: 1-55938-872-2

conflicting interests. Their centrality has led some scholars to predict that the study of top executives will be the central research thrust in the strategic management field in the 1990s (Hambrick 1989).

Evidence from both academic studies and from the popular press underscore the criticality of top team decisions and actions in determining a firm's direction and success. In the academic literature, top managers' effects on such outcomes as innovation (Bantel and Jackson 1989), strategy type (Gupta and Govindarajan 1984), strategic change (Wiersema and Bantel 1992), strategic diversification (Michel and Hambrick 1992), and performance (Murray 1989) have been examined. Countless indications of critical executive decision making exist in the popular press; for example, GM management's decision to cut 150,000 jobs during the 1980s (twenty-five percent of jobs) (Northcraft, Griffith, and Shalley 1992) will have a very substantial impact on its future success.

While we can observe that top management teams vary widely in their effectiveness, an important issue is understanding what characteristics and processes differentiate top management teams. Research in this area has included a focus on managerial characteristics, both personality (e.g., Miller and Toulouse 1986) and demographic (e.g., Bantel and Jackson 1989). Other studies have examined power (Hambrick 1981), political processes (Eisenhardt and Bourgeois 1988), managerial values (Hage and Dewar 1973), and roles (Mintzberg 1973). While such a broad-brushed research stream has many advantages, including the breadth of questions and issues addressed and quicker advancement of knowledge, it can also be somewhat disjointed, inhibiting comparability of findings across studies.

Our focus is on those attributes of teams that relate to their strategic decision-making process, and in turn, strategic actions and outcomes. We conceptualize top management teams as the key decision-making unit in an organization, processing information from internal and external sources to make decisions that affect firm strategy and performance, consistent with Mintzberg (1979). As the strategic decision-making process is complex, unstructured, and ambiguous (Mintzberg, Raisinghani, and Theoret 1976), managerial perceptions and interpretations play an important role (Dutton and Duncan 1987). To help us better understand such perceptions and interpretations, we focus on the structure of the top team, which is closely aligned with fundamental elements of team process (Keck 1990), including the strategic decision-making process. Our attention here is on both the structure of top management teams, defined as the central characteristics of which the senior management group is composed, and its antecedents.

The purpose of this paper is twofold: first, to provide insight for practicing managers on top team structural attributes that have an impact on team strategic effectiveness and on the major determinants of these attributes; and second, to motivate research in this area. We accomplish this by (1) conceptualizing top management team structure in a manner that lends insight into team decision process and (2) developing a model of the determinants of top team structure. By focusing on a limited, but central, set of top team structural dimensions, our goal is to provide focus to a relatively unfocused area of inquiry for both managers and researchers.

In the following section, we discuss the major attributes of top management team structure and their relationship to decision making. Next, we outline important antecedents and present several propositions for each to guide subsequent research. We conclude by discussing implications for practicing managers and, in addition, several issues that investigators should consider in testing these propositions.

TOP MANAGEMENT TEAM STRUCTURE AND STRATEGIC DECISION MAKING

While the top management team is theoretically defined as the group at the apex of an organization that is responsible for strategic decision making and subsequent organization performance, there is no one generally accepted operational definition. Hence, we propose the following: all inside board members. Inside board membership represents an absolute cutoff point between top managers and other managers that is analogous across firms and industries. Membership on the board is a formal, objective indicator of status; it effectively signifies membership in the "inner circle" (Thompson 1967) of top managers. Some support for this view is provided in recent work by Finkelstein (1988), who found inside board members (excluding the CEO) were judged more powerful than other top managers (at the senior vice-president level and higher).

The key dimensions of the team are those characteristics that are most likely to affect its strategic decision-making process and, hence, strategic direction and firm performance. In this paper, top team structure is defined in terms of demographic mean characteristics, demographic diversity, and team size. While alternative conceptualizations of top teams can be envisioned (for example, such psychological attributes as personality could be emphasized), the one proposed here is based on a concern with the top management team as the key strategic decision-making body of the firm. Each of the attributes links directly to the team's strategic decision-making process characteristics, as described subsequently. Further, this approach highlights enduring attributes of teams that enhances comparability across studies, facilitates empirical work, and is consistent with recent literature (e.g., Pfeffer 1983).

Team Decision-Making Process

The nature of the strategic decision-making process can be understood by a model proposed by Hambrick and Mason (1984), which draws from the Carnegie School behavioral perspective (Cyert and March 1963; March and Simon 1958). The model proposes that decision making of this type is a perceptual process in which individuals' cognitive bases serve as the foundation. The cognitive base, defined by March and Simon (1958), is that person's: "(1) knowledge or assumptions about future events, (2) knowledge of alternatives, and (3) knowledge of consequences attached to alternatives" (Hambrick and Mason 1984, p. 195). Starting with the cognitive base as input, a perceptual process occurs that proceeds in the following sequence: first, the field of vision is limited; second, selective perception occurs; and third, interpretations are made that determine managerial attitudes, perceptions, perspectives, and decision choices (Hambrick and Snow 1977). Hence, an understanding of the cognitive bases of team members will provide insight into the team's decision-making process and outcomes.

Such insight results from not only understanding the nature of the cognitive bases represented by team members, but also the variety of cognitive bases, or the cognitive diversity, of the team. Cognitive diversity represents the breadth of information and perspective at play among team members. Such diversity suggests a variety in the team's assumptions, information, and interpretations and has been shown to lead to more effective problem solving when the decision is complex and nonroutine (Filley, House, and Kerr 1976; Shaw 1976; Wanous and Youtz 1986). Diverse teams are more likely to have members challenge each other (Hoffman and Maier 1961), thus overcoming the "groupthink" phenomenon. On the other hand, diversity has been found to decrease communication frequency and quality (Wagner, Pfeffer, and O'Reilly 1984; O'Reilly, Caldwell, and Barnett 1989; Zenger and Lawrence 1989); such teams are also more likely to have problems achieving cohesiveness (Zander 1977) and reaching consensus (Hoffman and Maier 1961).

In summary, both the nature of the cognitive bases represented by team members and the team's cognitive diversity provide insight into the team's internal strategic decision-making process which, in turn, will influence decision outcomes. Three aspects of team structure, demographic mean characteristics, demographic diversity, and team size, reflect team cognitive bases and diversity.

Demographic Mean Characteristics and Diversity

The demographic composition of a top management team profiles the demographic characteristics, such as age and tenure, of the key decision makers in a firm. Top team demographics have been the focus of a variety of studies that relate to such outcomes as team turnover (Wagner, Pfeffer, and O'Reilly 1984; Jackson, Brett, Sessa, Cooper, Julin, and Peyronnin 1991), organizational innovation (Bantel and Jackson 1989), firm performance (Murray 1989), strategic decision processes (Frederickson and Iaquinto 1989), strategic change (Wiersema and Bantel 1992), and strategic diversification (Michel and Hambrick 1992).

Following Hambrick and Mason (1984), demographic characteristics are indicative of top managers' cognitive bases, attitudes, values, and expectations. Both the mean and distributional properties of these variables are important. Mean properties refer to the mean level (e.g., mean tenure) or proportion (e.g., percentage of top managers with marketing functional backgrounds) within the team on the demographic variable; distributional properties indicate the social differentiation or heterogeneity (e.g., variation in tenure) of the team. Both types of properties are consistent with Pfeffer's (1983) discussion of organization demography and both have been studied in earlier work in this area.

Demographics reflect a variety of perceptual tendencies critical to top team decision-making processes and outcomes. The mean demographic characteristics of the team, for example average age and tenure, are indicative of managers' cognitive bases which include such attitudes and perspectives as the approach to risk-taking (Vroom and Pahl 1971) and willingness to undergo strategic change (Wiersema and Bantel 1992). The demographic distributional characteristics represent the cognitive diversity described earlier. Hence, demographic diversity reflects the breadth of information gathering, communication frequency and quality (Wagner, Pfeffer, and O'Reilly 1984; O'Reilly, Caldwell, and Barnett 1989; Zenger and Lawrence 1989), creativity (Wanous and Youtz 1986), and cohesiveness (Zander 1977) that affect the ability of the team to reach consensus (Hoffman and Maier 1961). As a result, demographic characteristics are seen to be a central element of top management team structure. Based on their theoretical importance, the major demographic characteristics that we discuss are age, tenure in the organization, tenure in the industry, and functional background.

Team Size

The size of the top management team is an essential element of group structure. It defines the boundaries of the team, designating who is "in" and

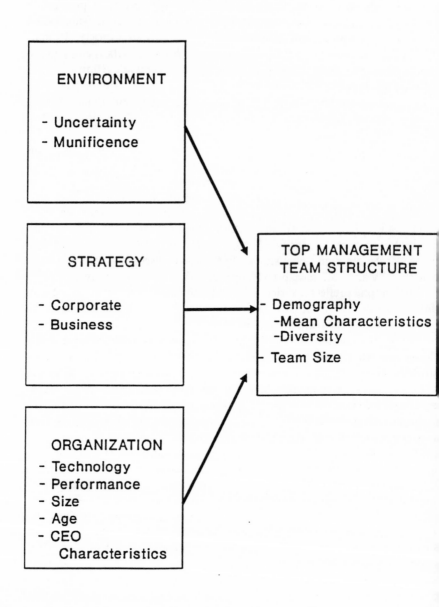

Figure 1.

who is "out." Team size is important because of its relatively straightforward association with cognitive diversity. Increases in team size will be associated with more diverse perspectives represented on the team as the elaboration of team structure (Meyer 1972) occurs. Further, increased differentiation of perspective (Dearborn and Simon 1958) is expected with increased team size; a commensurate increase in demographic differentiation (or diversity) is likely. These effects of increased team size are particularly true for small teams (Bantel and Jackson 1989).

As indicated earlier, the cognitive diversity of the team is of interest in that it indicates the breadth of information and perspective at play in the decision-making process. It further suggests the nature of the team's communication processes in that low diversity is related to more effective communication (Wagner, Pfeffer, and O'Reilly 1984; O'Reilly, Caldwell, and Barnett 1989; Zenger and Lawrence 1989). As a result of this increased differentiation, and because larger teams are also expected to show less cohesiveness (Shaw 1976), team size will also be negatively related to the tendency of the team to reach consensus and decisions will often take longer to reach (Hoffman and Maier 1961). Larger teams may be more effective problem solvers, on the other hand, especially in the complex and nonroutine decision situations that characterize strategic decision making (Filley, House, and Kerr 1976; Wanous and Youtz 1986).

In summary, a general model of the linkages between top management team structural attributes and various aspects of the team decision process, which in turn are linked to decision outcomes, is depicted in Figure 1. The structural attribute of demographic mean characteristics links to the nature of the cognitive bases; demographic diversity and team size relate to the team's breadth of information/perspective, communication processes, and the ability of the team to reach consensus. These aspects of team decision process, in turn, are expected to influence strategic decision outcomes. In the following section, we discuss the determinants of team structural attributes.

DETERMINANTS OF TOP MANAGEMENT TEAM STRUCTURE

This section examines the major determinants of top management team structure. Top team structure is derived from three sets of factors familiar to organizational researchers: environmental, strategic, and organizational. These are emphasized as they represent traditionally powerful influences on organizations and organizational behavior. We discuss several antecedents within each set, and offer propositions, depicted in Figure 2.

Determinants

Top Management Team Structure

Demographic Mean Characteristics

Environment[a]	Age	Tenure	Industry Tenure	Functional Background	Demographic Diversity	Team Size
Uncertainty - Instability	L[b]	L		External[c]	Heterogeneity[d] (A,T,I)	H
- Complexity		H		External	Heterogeneity (A,T,I)	H
Munificence	L	L[e]	L	External		H
Strategy						
Corporate - Level of Diversification		L	L	Internal; also Law		L
Business - "Internally Directed"	H	H		Internal	Homogeneity (A,T)	H
Organization						
Technological Complexity				External; also Personnel	Heterogeneity (A,T,I)	H
Performance		H			Homogeneity (T)	H
Size and Age	L	L[e]			Homogeneity (A,T) Heterogeneity (F)	H
CEO						
Tenure		H			Homogeneity (A,T,F)	H
Age					Homogeneity (A,T,F)	
Education Level					Heterogeneity (A,T,F)	

[a] For determinants, higher levels are assumed to associate with team structural attributes unless otherwise noted.
[b] L = low mean on attribute; H = high mean
[c] External functions include marketing, sales, and product R&D; internal functions include production/operations, accounting, finance, and process R&D
[d] Heterogeneity on A = age, T = tenure, I = industry tenure, F = functional background
[e] Associated with both very high and very low levels of determinant

Figure 2.

Environment

Environmental influence on organizations is a major theme in organization theory; we draw on several theories to discuss how organizational environments affect the structure of top management teams. Two dimensions of the environment will be considered: uncertainty (including instability and complexity) and munificence. These dimensions were selected as they are among the most critical in ensuring the firm's survival (Aldrich 1979; Dess and Beard 1984; Duncan 1972) and are, hence, expected to have an important influence on top team structure.

Environmental Uncertainty

The degree of uncertainty in an organization's environment is an important construct in organization theory; we suggest that it also has important effects on the structure of top management teams. Organizations need top managers who can cope with environmental uncertainty (Thompson 1967). Hence, we would expect to see top management teams reflecting this dominant environmental need (Hambrick 1981).

Following Emery and Trist (1965), Thompson (1967), and Terreberry (1968), two dimensions characterize environmental uncertainty: simple-complex and static-dynamic. The first dimension relates to the range of and heterogeneity in the number of environmental factors that impinge upon the organization and strategic decision making and which, therefore, require monitoring (Child 1972; Duncan 1972); the second dimension refers to stability, or the rate of change, in those factors (Duncan 1972). As these dimensions are both conceptually and empirically distinct (Dess and Beard 1984), they will be treated separately in the discussion of their impact on top management team structure.

Environmental Instability. Environmental instability affects the demographic composition of the top team in several ways. Because such environments create risk (Galbraith 1973), top managers may be younger, with short tenures. For example, several studies have found that environmental instability is associated with CEO turnover (Osborn, Jauch, Martin, and Glueck 1981; Pfeffer and Leblebici 1973; Pfeffer and Salancik 1978). In a similar vein, Harrison, Torres, and Kukalis (1988) found that firms in industries with intermediate concentration (that face the highest competitive uncertainty [Pfeffer and Salancik 1978]) experienced the greatest top management team turnover.

Environmental instability may also influence the functional background of top executives. Firms faced with an unstable environment may feel pressure to focus on externally-oriented functions such as marketing and product development in order to cope (Miles and Snow 1978); internally-oriented functions such as production and accounting will likely provide the critical management information in more stable environments (Snow and Hrebiniak 1980). In a similar vein, Hambrick and Mason (1984) argue that for firms in turbulent industries, output functional experience (marketing, sales and product R&D) is necessary; throughput functional experience (production, process engineering, and accounting) is important in stable industries. In addition, Daft, Sormunen, and Parks (1988) found that scanning activity among executives increased in firms with high environmental instability, suggesting the importance of externally oriented functions. Thompson (1967) makes a similar point about top teams directing managerial attention externally as they attempt to reduce the environmental uncertainty facing the firm.

A summary of the propositions described here is found in Figure 2.

Environmental Complexity. Environmental complexity is another important antecedent of top team structure. Gupta and Govindarajan (1984) found that industry familiarity was important for executives in complex environments, suggesting that industry tenures may be longer. The distributional properties of top team demographic composition are also likely to be influenced by environmental complexity. Organizations in complex environments are confronted with conflicting demands from multiple constituencies, promoting structural differentiation (Thompson 1967; Pfeffer and Salancik 1978). Such differentiation is also reflected in the top management team (Lawrence and Lorsch 1967; Arrow 1974), as members attempt to monitor the diversity and complexity of their environment. Tushman and Keck (1991) suggest that a competitive environment (which entails considerable complexity) requires a differentiated top team, particularly with respect to functional background. We suggest that the top management team may be differentiated on a number of demographic characteristics. Firms operating in complex environments often face ill-defined and novel problem solving situations that require cognitively diverse and heterogeneous teams (Janis 1972; Hambrick and Mason 1984). In simpler, less complex environments, such diversity is not required; it may also not even be desirable as diversity can create such dysfunctional group process effects as poor communication (Zenger and Lawrence 1989) and low group cohesion and integration (Ebadi and Utterback 1984; O'Reilly, Caldwell, and Barnett 1989).

Environmental complexity is expected to influence the functional background of top executives in a manner similar to environmental instability. Managerial attention will be focused externally in attempts to understand and manage the complex environment, consistent with Thompson (1967). The study by Daft, Sormunen, and Parks (1988) described earlier also found that high environmental complexity increased external scanning activity among executives, suggesting that externally oriented functions will be important.

Both dimensions of environmental uncertainty affect team size as well. As instability increases, top teams expand to take in more members who can provide needed competencies (Tushman and Keck 1991). The multiple constituencies that firms typically face in complex environments present top teams with management challenges on several fronts, bolstering team size. As Thompson (1967) has argued, the more sources of uncertainty for the organization, from either instability or complexity, the larger the size of the dominant coalition. While top managers might be aware that larger teams take longer to reach consensus on decisions (Hoffman and Maier 1961), they might be willing to add executives as the benefits of a larger team would seem to outweigh this concern. (See Figure 2 for a summary of this discussion.)

Environmental Munificence

Environmental munificence refers to the extent to which the environment supports sustained growth (Starbuck 1976). As such, it is quite similar to Aldrich's (1979) concept of environmental capacity (Dess and Beard 1984) which emphasizes resource availability. Organizational growth and stability (Dess and Beard 1984) and the creation of slack (Cyert and March 1963) are associated with environmental munificence. Hence, munificent environments offer significant discretion to top management teams (Hambrick and Finkelstein 1987), as they are relatively unfettered by external demands and are an important influence on the structure of top teams.

Work that has examined the relationship between industry growth (an important indicator of munificence [Dess and Beard 1984]) and managerial turnover has found mixed results. On the one hand, Pfeffer (1983) has argued that firms in high growth industries tend to raid each other of executive talent to meet staffing goals and to reduce the risk of hiring a person new to the industry. His study of the semiconductor industry (Pfeffer 1980) confirmed the existence of short tenures in a rapidly growing industry. Similarly, Pfeffer and Leblebici (1973) found a positive association between industry growth and the number of job changes by executives. Finally, Harris (1979) found that slow industry growth reduced executive mobility, leaving firms with older top managers with long tenures.

Conversely, Wiersema and Bantel (1993) found a negative relationship between munificence (measured as growth in industry sales) and turnover within the top team in their Fortune 500 study. These authors posit that a lack of munificence creates a threatening environment, leading to dysfunctional strategic decision-making processes and inappropriate strategic decisions; team members leave as a result.

These inconsistent results might be attributable to the fact that Pfeffer and colleagues focused on rapidly growing industries in which competition for managerial talent is fierce; Wiersema and Bantel (1993) focused on Fortune 500 firms where high munificence indicates steady and stable growth and low munificence characterizes stagnant industries. Hence, it might be that an inverted U relationship exists in which both very high and very low munificence contributes to managerial turnover and, therefore, to lower average tenures; intermediate levels of munificence represent stability and longer tenures. This relationship is depicted in Figure 2.

Very strong industry growth might have a tendency to attract younger managers to the industry who are more inclined to have a higher risk-taking attitude (Vroom and Pahl 1971). High growth might also necessitate hires from outside the industry as the pool of experienced managers cannot keep pace with the demand, shortening the average industry tenure of managers.

Environmental munificence also affects functional backgrounds in top teams. Top teams in munificent environments typically have the discretion to search for new opportunities (Hrebiniak and Joyce 1985). Those managers with expertise in scanning the environment for market opportunities and capitalizing on them will likely predominate, suggesting that marketing, sales, and product R&D are important functions. Conversely, when environments offer top teams little opportunity for growth, inertial tendencies expand and strategic changes are constricted (Hambrick and Finkelstein 1987). An environment lacking in munificence might signal an industry in the later stages of its life cycle, a period during which firms will often attempt to maximize profits while minimizing investments in such areas as new markets and new production processes (Hofer 1975). Thus, more constrained environments challenge top managers to make more effective use of their resources, which production, accounting, and process R&D all emphasize.

Team size is also related to environmental munificence. Organizations in munificent environments often generate significant slack resources (Cyert and March 1963), which often takes the form of increases in staff. For example, Williamson (1963) has argued that firms with slack resources tend to emphasize such "pecuniary" benefits as enlarged executive staffs. Those firms in stagnant industries, by contrast, will have a tendency to focus on cost containment (Hofer 1975), including reduction of staff at all levels including the top team. (This discussion is summarized in Figure 2.)

Strategy

The strategy a firm adopts has an important effect on many aspects of its organization. For example, diversified firms often institute a multidivisional structure (Chandler 1962). In a similar vein, the structure of top management teams is also expected to be affected by firm strategy. Gupta (1984) points out that strategic contexts differ in terms of the knowledge, values, skills, and personality orientation required of key executives. Hence, there is a strong a priori case for expecting an association between strategy and top management team structure. The following sections discuss the association between team structure and both corporate and business strategies.

Corporate Strategy

A central element of a firm's corporate strategy is the degree of diversification of its businesses (Rumelt 1974). There is considerable evidence to suggest that the degree of firm diversification affects the structure of top management teams, especially its demographic composition.

Managerial tenure in the organization and the industry appear to be influenced by the degree of firm diversification. Less diversified firms (with single business firms being least diversified, followed by vertically integrated firms, related diversifiers, and unrelated diversifiers) require intimate knowledge of product/market strategy and key competitive factors. As a result, both firm-specific and industry-specific knowledge are highly valued; thus, longer managerial tenures (in the organization and the industry) are likely. In contrast, managing the unrelated diversified firm has often been likened to managing a business portfolio (Rumelt 1974; Salter and Weinhold 1978; Gupta 1984); little industry familiarity is needed. As firm diversification becomes more unrelated, firm and industry tenure become less important (Gupta 1984) and multiple industry experience can be beneficial (Smith and White 1987).

Research on strategy implementation suggests that certain functional specializations are critical to the successful implementation of corporate strategy. Top managers in less diversified firms need to have in depth expertise in the management of their key business. Hence, they are more likely to have functional expertise in the core areas of marketing, R&D, and operations (Smith and White 1987), the latter especially characteristic of vertically integrated firms. In contrast, top managers of unrelated diversified firms are less involved in internal operations, operate in multiple environments, and are more externally-oriented (Lorsch and Allen 1973); they emphasize the portfolio management skills of accounting, finance, and law (Song 1982; Smith and White 1987).

The degree of diversification is also related to team size. Managing highly diversified firms, such as unrelated diversifiers, is similar to managing a financial portfolio (Berg 1969; Rumelt 1974). Top teams are more concerned with buying and selling businesses than with actively managing the operations of each business. Because the task of operating the businesses is delegated down to general managers at the divisional level, the corporate offices of highly diversified firms tend to be quite small (Pitts 1977). The less diversified firm, which is more likely to have a functional structure (Rumelt 1974), will have a larger top team that is typically involved in all functions of the firm's operations. (See Figure 2 for a summary of this discussion.)

Business Strategy

There have been several attempts in the literature to relate competitive strategies to the structure of top management teams, especially demographic composition (cf. Gupta 1984). In this section, we synthesize this work by describing two commonly cited strategic types and discuss their implications for top team structure. The first type, labeled "internally-directed strategy," is a composite of several strategies: harvest (Gupta and Govindarajan 1984), low-cost producer (Porter 1980), throughput (Hambrick and Mason 1984), and defender (Miles and Snow 1978). The second composite type is "externally-directed," and includes build (Gupta and Govindarajan 1984), differentiation (Porter 1980), output (Hambrick and Mason 1984), and prospector (Miles and Snow 1978) strategies. Although there are some differences among the strategies clustered in each group, their implications for top management teams are remarkably similar.

Firms with internally-directed strategies emphasize cost control, stability, and efficiency. Those with externally-directed strategies emphasize growth, search for new opportunities, and follow generally aggressive strategies. As such, internally-directed firms face less risk than do firms with externally-directed strategies (Gupta and Govindarajan 1984), suggesting that top managers may be older (Child 1974) and have longer tenures (Coffee 1988). For example, older managers, seeking to avoid risky situations (Carlson and Karlsson 1970; Vroom and Pahl 1971) that might threaten financial security, are less likely to lead firms with relatively risky strategies. In a similar vein, Chaganti and Sambharya (1987) found that outsiders were more common in prospectors than in defenders in a sample of tobacco firms.

As one of the primary thrusts of an externally directed strategy is growth, firms following internally-directed strategies by comparison are more stable and are less oriented to growth; as a result, promotion opportunities will be relatively more limited. As a result, younger and less-tenured managers may leave the firm (Pfeffer 1983), not only contributing to the longer managerial tenures and ages cited earlier, but also implying greater homogeneity in age and tenure at the top.

Many studies have found a link between functional backgrounds and business strategy. Typically, top teams with internally-directed strategies emphasize accounting, finance, and operations, while senior managers in externally-directed firms exhibit product R&D, marketing, and sales backgrounds. For example, accounting, finance, and production are important for harvest strategies; finance and production are the most important functional areas for defenders; and process R&D is key for low cost producers (Gupta 1984; Gupta and Govindarajan 1984; Miles and Snow 1978). In contrast, pursuit of a build strategy involves marketing and sales expertise; marketing and R&D backgrounds are most important for prospectors; while firms with differentiation strategies emphasize marketing and product R&D (Gupta and Govindarajan 1984; Miles and Snow 1978). The findings on prospectors are further supported by Hambrick (1982) who found that for two of three industries he studied, prospectors engaged in more environmental scanning than defenders did, consistent with such externally oriented functions as marketing. These results mirror those proposed by Hambrick and Mason (1984) for output and throughput strategies.

An organization's business level strategy also influences team size. Given the stability within firms following internally-directed strategies, we would expect to find increased bureaucratic momentum (Tushman and Romanelli 1985), including a gradual increase in team size (Pondy 1969). (This discussion is summarized in Figure 2.)

Organization

In addition to the role of environment and strategy, the structure of top management teams is also affected by organizational factors. We discuss five major influences: technology, performance, size, age, and CEO characteristics.

Technology

The technology of the firm is concerned largely with the throughput mechanisms employed by the organization to produce its products and services and has been conceived of as a causal influence on various aspects of organizational structure (e.g., Perrow 1970). In a similar vein, we develop its relationship to the top management team structure.

Technology plays a particularly important role in influencing functional backgrounds. Firms with complex technologies (e.g., small batch) tend to produce differentiated products that require strong R&D skills to cultivate. Selection and training of highly specialized personnel is often needed. And because the market for highly differentiated products may be small, strong marketing skills help to develop the firm's niche. As a result, researchers have found that firms with complex technologies emphasize R&D (Woodward 1965; Lawrence and Lorsch 1967), personnel (Perrow 1970), and marketing (Lawrence and Lorsch 1967) functional backgrounds. These findings are consistent with Hambrick and Mason's (1984) discussion of the importance of marketing and product R&D in differentiated industries.

Firms with less complex technologies (e.g., mass production) often produce commodity goods. These firms tend to be more concerned with price (Porter 1980) and may be larger than many differentiated firms serving specific niches. Such firms tend to emphasize production (Woodward 1965; Perrow 1970; Lawrence and Lorsch 1967) and general administration (Perrow 1970). These conclusions are in line with Hambrick and Mason's (1984) suggestion that firms in commodity industries, seeking efficiency in the transformation of inputs to outputs, value functional experience in production, accounting, and process R&D.

There is also some evidence to suggest that the complexity of the organization's technology affects the distributional properties of the top management team. Technological complexity creates important contingencies for the organization to manage (Thompson 1967). As the complexity of the organization's task increases, the composition of the top management team will reflect this complexity (Bower 1973; Cyert and March 1963) by being more diverse in backgrounds and experiences. We would expect this diversity to be most evident in members' age, and organization and industry tenure; such diversity would not be expected in functional backgrounds as a result of the argument stated earlier.

The size of the top management team will also be related to the complexity of the organization's technology. As described earlier, the top team will become more diverse in backgrounds and experience in order to monitor and control complex technology (Bower 1973; Cyert and March 1963). As a result, a larger number of top team members will be needed. And because the command of a complex technology can itself become an important source of competitive advantage (Von Glinow and Mohrman 1990), such control will be highly sought after. (See Figure 2 for a summary.)

Organizational Performance

Organizations that are performing well tend to have stable top management teams. Numerous studies have shown that good performance is associated with long managerial tenures, while poor performance is more likely to result in turnover of the CEO (Allen, Panian, and Lotz 1979; Gamson and Scotch 1964; Grusky 1963; Helmich 1980; Tushman, Virany, and Romanelli 1985) or top management team members (Wagner, Pfeffer, and O' Reilly 1984). Hence, we would expect firms with strong performance records to have top management teams with longer tenures. Further, given that such firms tend to make few changes (Tushman and Romanelli 1985), greater homogeneity in team tenures is expected as well.

Organizational performance will also influence team size. Firms that are performing well tend to generate excess slack resources (Miles and Cameron 1982) that contribute to the formation of larger top teams. (Figure 2 summarizes this discussion.)

Organizational Size and Age

Both organizational size and age affect the structure of top management teams in several ways. As size and age are highly correlated (Hannan and Freeman 1977), we expect them to be generally similar to each other with regard to their influence on the team. The effects of both on each aspect of team structure will be described where relevant.

Research on the relationship between organizational size/age and team tenure is somewhat mixed. Several studies have found a positive relationship between size and top management team turnover (Crain, Deaton, and Tollison 1977; Grusky 1961; Harrison, Torres, and Kukalis 1988; James and Soref 1981). This can be expected for two reasons. First, firm size is associated with both formalization (Grusky 1961) and decentralization (Harrison, Torres, and Kukalis 1988); Pfeffer and Moore (1980) point out that these characteristics allow firms to tolerate administrative succession better. Second, Harrison, Torres, and Kukalis (1988) argue that senior managers in large firms are highly visible within the industry, leading to a greater number of job offers. Hence, one view is that larger firms have top teams with shorter tenures.

On the other hand, younger and smaller firms could also be expected to have high team turnover. Organizations adapt to their environments over time. For younger firms, this process can be problematic, as they have no history to rely on and may be encountering elements of their task environment for the first time (Tushman and Keck 1991). As a result, young firms are much more unstable than are more established firms (Stinchcombe 1965). Top management teams of young firms reflect this instability by being unsettled, with significant compositional change occuring. As a result, young firms often exhibit short managerial tenures.

These inconsistent results can perhaps be reconciled by positing an inverted U relationship between firm size/age and team tenure. At the extremes of very small/young firms and very large/old firms, there is a higher tendency for top team turnover; at intermediate levels of size and age, we would expect high stability and the concomitant low turnover and high tenure.

The age of top managers is expected to increase with the size and age of the firm. As entrepreneurs tend to be younger (older managers tend to be more established, with more to lose by starting or joining a new venture), younger and smaller firms tend to have younger top managers. Support for this was found by Bantel (1993) who found a significant correlation between average team member age and firm size.

Larger and older firms are also expected to exhibit homogeneity on age and tenure. As such firms have a tendency to resist change, both within the organization and within the team, greater homogeneity in age and tenure is promoted (Pondy 1969; Tushman and Keck 1991). As executives leave the largest and oldest firms as described earlier, inside succession is more likely as large candidate pools consisting of experienced executives exist (Dalton and Kesner 1983; Furtado and Rozeff 1987; Helmich 1975, 1977; Pfeffer and Moore 1980; Reinganum 1985) and the entrenched power structure supports continuity. Consistent with Hannan and Freeman's (1977) argument concerning structural inertia, we would expect the successor to be similar to his or her predecessor and to initiate little change, decreasing the variation in age and tenures. Bantel (1993) found empirical support for a relationship between organization size and homogeneity on both age and tenure.

As the size and age of the firm increases, structures also become more elaborated (Meyer 1972) and differentiated (Blau 1970), leading to a wider variety of beliefs about cause-effect relationships (Dearborn and Simon 1958). Larger firms have more demands placed on them by a variety of interests (Pfeffer and Salancik 1978). Hence, we would expect larger firms to exhibit greater functional heterogeneity within the top management team, easier to achieve as the team grows larger as willl be described. This expectation is supported by the work of Bantel and Jackson (1989).

Top team size is expected to exhibit a positive relationship with both organization size and age. Larger and older firms are generally more complex and bureaucratic, and typically operate in numerous domains, buoying administrative intensity (Hambrick and Finkelstein 1987). With limits to the span of control, more top managers will be needed and teams will be larger. Further, we would expect team size to increase in larger and older firms as bureaucratic momentum gains. In contrast, most strategic decisions in new, small firms are made by the chief executive, who presides over a small managerial hierarchy (Mintzberg 1979). Younger firms are also more constrained for resources, reducing team size. Support for a relationship between organization size and team size was found by both Bantel and Jackson (1989) and Wiersema and Bantel (1992). (See Figure 2 for a summary.)

CEO Characteristics

As the most powerful member of the top management team, the chief executive officer clearly plays a central role in its functioning. However, the CEO's influence extends beyond his or her membership in the team; just as the other environmental, strategic, and organizational factors discussed earlier affect the type of team that is formed, so too does leadership. Although modeling CEO characteristics as both an antecedent of top teams and a part of the top team does present an analytical problem, the discussion and propositions that follow are intended to inform theory and are testable. The propositions that follow pertain to the influence of the CEO on the structure of the remaining subgroup, that is the cadre of top managers aside from the CEO.

Although CEO preferences may be exhibited in myriad ways, we limit our discussion to the influence of readily observable demographic characteristics as predictive of team structure. We focus here on those characteristics for which previous research suggests a potential linkage with team structure—CEO tenure in the organization, age, and education.

CEO Tenure. There is considerable evidence to suggest that long CEO tenures are associated with stability and continuity (Katz 1981), both in the organization and the top management team (Carlson 1962; Kotin and Sharif 1976). In contrast, short-tenured CEOs, and particularly those who have just succeeded to the job, tend to disrupt existing patterns of strategy and structure (Carlson 1972; Gabarro 1987; Helmich and Brown 1972). Managers associated with the departing CEO's team may leave; over time, new executives are promoted or recruited to take their place (Tushman and Keck 1991). Hence, we would expect CEO tenure to be positively associated with top managerial tenure and tenure homogeneity within the team. In addition, because long-term processes of socialization and filtering reduce the variance in the population of promotion candidates in a firm (March and March 1977), tenure is expected to promote greater homogeneity in age and functional backgrounds as well.

Enduring CEO tenures enhance organizational stability and continuity over time (Katz 1981). Accompanying bureaucratic momentum is expected to gradually expand the size of the top management team as well. In addition, a CEO's power may become institutionalized over time (Pfeffer 1981), facilitating expansion of the size of the team as a pecuniary benefit (Williamson 1963). (See Figure 2 for a summary.)

CEO Age. CEO age is associated with less risk taking (Vroom and Pahl 1971), less flexibility (Taylor 1975), and greater resistance to change (Child 1974). Older executives tend to be more committed to the status quo (Alutto and Hrebiniak 1975; Stevens, Beyer, and Trice 1978), in part because financial and career security gain in importance as managers age (Hambrick and Mason 1984). Any change that may jeopardize their positions is likely to be avoided. Hence, older CEOs are likely to seek the certainty and continuity that homogeneous top teams can provide. (Figure 2 summarizes this discussion.)

CEO Education Level. Highly educated chief executives may prefer more heterogeneous teams. Hambrick and Mason (1984) point out that higher education for managers, particularly those with a business discipline, is likely to have a positive influence on administrative complexity and sophistication, consistent with team heterogeneity. Further, a consistent finding in the innovation literature is that the greater an individual's education, the more receptive he or she is to innovation (Bantel and Jackson 1989; Kimberly and Evanisko 1981; Rogers and Shoemaker 1971). Hence, highly educated CEOs may try to structure the top management team to enhance innovativeness. Because top managers typically confront complex, non-routine problems (Mintzberg, Raisinghani, and Theoret 1976), the most effective groups are those composed of individuals having a variety of skills, knowledge, abilities, and experiences (Shaw 1976; Wanous and Youtz 1986). CEOs seeking innovativeness at the top will want to promote the cognitive diversity that comes from having a heterogeneous top team. As a result, we would expect CEO education to be positively associated with team heterogeneity (in age, tenure, and functional background). (See Figure 2 for a summary.)

CONCLUSIONS

Given the importance of the top management team to organizational outcomes, it seems important to gain a better understanding of the ways in which teams differ. Our approach has been to isolate the major attributes of top team structure that provide insight to strategic decision making and organizational performance. We discussed the major antecedents of these dimensions of top team structure, incorporating environmental, strategic, and organizational factors. While there may be some additional influences that warrant consideration, we hope to have at least outlined the major determining characteristics that are central to strategic management and organization theory.

Implications for Human Resource Managers

A major goal of this work was to offer insight to top executives, including those responsible for human resource concerns, on the key structural attributes of top management teams that link with their strategic effectiveness, and on the contextual influences within which top management teams are structured.

We hope we have drawn attention to the importance of demographics, both mean characteristics and their diversity, in determining the nature of decision-making interactions among top team members. In particular, we have emphasized the importance of age, organizational tenure, industry tenure, and functional background as indicators of the strategic decision-making perspective of executives. As indicated in our theoretical development, such specific demographic characteristics and their diversity can be linked to such strategy-related orientations as risk-taking, commitment to the status quo, functional viewpoint, tendency to challenge other members' viewpoints, and creativity. By being aware of these tendencies, there might be some opportunity for human resource executives to mold top team composition, through such approaches as succession planning or hiring practices, to achieve desired strategic decision-making outcomes.

While some deliberate molding of top teams might be feasible, we hope we have also alerted executives to other influences on the team's structural attributes that need to be considered. An understanding of the nature of a structural antecedent will serve as a starting point for determining both the likely and ideal team attributes. In some cases, the associations developed in the paper are purely descriptive, in others they are prescriptive; the prescriptive links suggest a positive effect on strategic decision making if the appropriate match exists. Prescriptive links between the environment and team structure, for example, are the following: if the environment is characterized by instability, then powerful top managers should have externally oriented functional backgrounds; complex environments are better managed by team members with diverse backgrounds. Similar conclusions can be drawn about other prescriptive links between the antecedents described and team structure.

Implications for Researchers

Another key goal of this paper was to suggest propositions to encourage future research. Those interested in investigating these questions should be sensitive to at least two issues. First, as we have discussed, most of the previous work on top management teams has examined their consequences. This suggests that special care will be needed to disentangle causal direction when testing some of our propositions, such as those linking strategic and organizational factors to top teams. Researchers will need to acknowledge reverse causality and design their studies to minimize its impact. Building in time lags in tested models, using change variables, or modeling a system of simultaneous equations are some of the ways to address this issue. Although the problem of causality is one that many scholars have grappled with (Hambrick and Mason 1984; Gupta and Govindarajan 1984), adequate resolution requires careful study design and interpretation of results. Finally, our model of structural antecedents offered relatively straightforward propositions without considering possible contingency linkages. It is apparent, however, that some antecedents may be more predictive under certain conditions. For example, Snow and Hrebiniak (1980), investigating the association between strategy and functional background, found that patterns varied by industry. In addition, some of the factors we discuss may interact in ways that we have not addressed. By not explicitly discussing these issues, we have necessarily limited the applicability of our model. However, given that this is one of the first attempts to model the antecedents of top management teams, such a limitation may be warranted. It is hoped that by testing some of our propositions, other researchers will be able to refine and extend our framework, and help to build a more complex model of top management teams.

ACKNOWLEDGMENT

We would like to thank Don Hambrick, Sara Keck, and Nandini Rajagopalan for helpful comments on an earlier draft of this paper.

REFERENCES

Aldrich, H. 1979. *Organizations and Environments*. Englewood Cliffs, NJ: Prentice Hall.
Allen, M. P. , S. K. Panian, and R. E. Lotz. 1979. "Managerial Succession and Organizational Performance: A Recalcitrant Problem Revisited." *Administrative Science Quarterly* 24: 167-180.
Alutto, J. and L. Hrebiniak. 1975. "Research on Commitment to Employing Organizations: Preliminary Fndings on a Study of Managers Graduating from Engineering and MBA programs." Paper presented at the meeting of the Academy of Management, New Orleans.
Arrow, K. 1974. *The Limits of Organization*. New York: Norton.
Bantel, K. 1992. "Strategic Clarity in Banking: The Role of Top Management Team Demography." *Psychological Reports* 73:1187-1201.
Bantel, K. and S. Jackson. 1989. "Top Management and Innovations in Banking: Does the Composition of the Top Team Make a Difference?" *Strategic Management Journal* 10 Special issue: 107-124.

Berg, N. A. 1969. "What's Different About Conglomerate Management?" *Harvard Business Review* 476: 112-120.

Blau, P. A. 1970. "Formal Theory of Differention in Organizations." *American Sociological Review* 35: 201-218.

Bower, J. L. 1973. *Managing the Resource Allocation Process*. Homewood, IL: Irwin.

Carlson, R. 1962. *Executive Succession and Organizational Change*. Danville, IL: Interstate Printer and Publishers.

————. 1972. *School Superintendents: Careers and Performance*. Columbus, OH: Merrill.

Carlson, R. K. and Karlsson. 1970. "Age, Cohorts, and the Generation of Generations." *American Sociological Review* 35: 710-718.

Chaganti, R. and R. Sambharya. 1987. "Strategic Orientation and Characteristics of Upper Management." *Strategic Management Journal* 8: 393-401.

Chandler, A. D. 1962. *Strategy and Structure*. Boston: MIT Press.

Child, J. 1972. "Organization Structure, Environment, and Performance: The Role of Strategic Choice." *Sociology* 6: 1-22.

————. 1974. "Managerial and Organizational Factors Associated with Company Performance." *Journal of Management Studies* 11: 13-27.

Coffee, J. C. 1988. "Shareholders Versus Managers: The Strain in the Corporate Web." In *Knights, Raiders, and Targets*, edited by J. C. Coffee, L. Lowenstein, and S. Rose-Ackerman. New York: Oxford University Press.

Crain, W. M., T. Deaton, and R. Tollison. 1977. "On the Survival of Corporate Executives." *Southern Economic Journal* 43: 1372-1376.

Cyert, R. and J. March. 1963. *A Behavioral Theory of the Firm*. New Jersey: Prentice-Hall.

Daft, R., J. Sormunen, and D. Parks. 1988. "Chief Executive Scanning, Environmental Characteristics, and Company Performance: An Empirical Study." *Strategic Management Journal* 9: 123-139.

Dalton, D. and I. Kessner. 1983. "Inside/Outside Succession and Organizational Size: The Pragmatics of Executive Replacement." *Academy of Management Journal* 26: 49-762.

Dearborn, D. C. and H. A. Simon. 1958. "Selective Perception: A Note on the Department Identification of Executives." *Sociometry* 21: 140-144.

Dess, G. and D. Beard. 1984. "Dimensions of Organizational Task Environments." *Administrative Science Quarterly* 19: 52-73.

Duncan, R. 1972. "Characteristics of Organizational Environments and Perceived Environmental Uncertainty." *Administrative Science Quarterly* 17: 313-327.

Dutton, J. and R. Duncan. 1987. "The Creation of Momentum for Change Through the Process of Strategic Issue Diagnosis." *Strategic Management Journal* 8: 279-296.

Ebadi, Y. and J. Utterback. 1984. "The Effects of Communication on Technological Innovation." *Management Science* 305: 572-585.

Eisenhardt, K. M. and L. J. Bourgeois. 1988. "Politics of Strategic Decision Making in High Velocity Environments: Toward a Midrange Theory." *Academy of Management Journal* 31: 737-770.

Emery, F. and E. Trist. 1965. "The Causal Texture of Organizational Environments." *Human Relations* 18: 21-32.

Filley, A., R. House, and S. Kerr. 1976. *Managerial Process and Organizational Behavior*. Glenview, IL: Scott Foresman.

Finkelstein, S. 1988. "Managerial Orientations and Organizational Outcomes: The Moderating Roles of Managerial Discretion and Power." Unpublished doctoral dissertation, Columbia University, New York.

Frederickson, J. and A. Iaquinto. 1989. "Inertia and Creeping Rationality in Strategic Decision Processes." *Academy of Management Journal* 32: 516-542.

Furtado, E. P. and M. S. Rozeff. 1987. "The Wealth Effects of Company Initiated Management Changes." *Journal of Financial Economics* 18: 147-160.

Gabarro, J. 1987. *Dynamics of Taking Charge*. Boston: Harvard Business School Press.

Galbraith, J. 1973. *Designing Complex Organizations*. Reading, MA: Addison-Wesley.

————. 1967. *The New Industrial State*. Boston: Houghton Mifflin.

Gamson, W. A. and N. Scotch. 1964. "Scapegoating in Baseball." *American Journal of Sociology* 70: 69-72.

Grusky, O. 1961. "Corporate Size, Bureaucratization, and Managerial Succession." *American Journal of Sociology* 67: 261-269.

————. 1963. "Managerial Succession and Organizational Effectiveness." *American Journal of Sociology* 67: 261-269.

Gupta, A. K. 1984. "Contingency Linkages Between Strategy and General Manager Characteristics: A Conceptual Examination." *Academy of Management Review* 9: 399-412.

Gupta, A. K. and V. Govindarajan. 1984. "Business Unit Strategy, Managerial Characteristics, and Business Unit Effectiveness at Strategy Implementation." *Academy of Management Journal* 27: 25-41.

Hage, J. and R. Dewar. 1973. "Elite Values Versus Organizational Structure in Predicting Innovation." *Administrative Science Quarterly* 18: 279-290.

Hambrick, D. C. 1981. "Environment, Strategy, and Power Within Top Management Teams." *Administrative Science Quarterly* 26: 253-276.

————. 1982. "Environmental Scanning and Organizational Strategy." *Strategic Management Journal* 3: 159-174.

————. 1989. "Guest Editor's Introduction: Putting Top Managers Back into the Strategy Picture." *Strategic Management Journal* 10: 5-15.

Hambrick, D. C. and S. Finkelstein. 1987. "Managerial Discretion: A Bridge between Polar Views of Organizational Outcomes." Pp. 369-406 in *Research in Organizational Behavior* (Volume 9), edited by L. L. Cummings and B. Staw. Greenwich, CT: JAI Press.

Hambrick, D. C. and P. A. Mason. 1984. "Upper Echelons: The Organization as a Reflection of its Top Managers." *Academy of Management Review* 9: 193-206.

Hambrick, D. C. and C. Snow. 1977. "A Contextual Model of Strategic Decision Making in Organizations." *Academy of Management Proceedings*: 109-112.

Hannan, M. and J. Freeman. 1977. "Structural Inertia and Organizational Change." *American Sociological Review* 49: 149-164.

Harris, R. G. 1979. "The Potential Effects of Deregulation upon Corporate Structure, Merger Behavior and Organizational Relations in the Rail Freight Industry." Draft Report. Washington, DC: Public Interest Economics Center.

Harrison, J. R., D. L. Torres, and S. Kukalis. 1988. "The Changing of the Guard: Turnover and Structural Change in the Top-management Position." *Administrative Science Quarterly* 33: 211-232.

Helmich, D. 1975. "Corporate Succession: An Examination." *Academy of Management Journal* 3: 329-441.

————. 1977. "Executive Succession in the Corporate Organization: A Current Integration." *Academy of Management Review* 2: 252-266.

————. 1980. "Board Size Variation and Rates of Succession in the Corporate Presidency." *Journal of Business Research* 8: 51-63.

Helmich, D. and W. Brown. 1972. "Successor Type and Organizational Change in the Corporate Enterprise." *Administrative Science Quarterly* 17: 371-381.

Hofer, C. 1975. "Toward a Contingency Theory of Business Strategy." *Academy of Management Journal* 187: 84-810.

Hoffman, R. and N. Maier. 1961. "Quality and Acceptance of Problem Solutions by Members of Homogeneous and Heterogeneous Groups." *Journal of Abnormal and Social Psychology* 62: 401-407.

Hrebiniak, L. and W. Joyce. 1985. "Organizational Adaptation: Strategic Choice and Environmental Determinism." *Administrative Science Quarterly* 30: 336-349.

Jackson, S. J. Brett, V. Sessa, D. Cooper, J. Julin, and K. Peyronnin. 1991. "Some Differences Make a Difference: Individual Dissimilarity and Group Heterogeneity as Correlates of Recruitment, Promotions and Turnover. *Journal of Applied Psychology* 76: 675-689.

James, D. R. and M. Soref. 1981. "Profit Constraints on Managerial Autonomy: Managerial Theory and the Unmaking of the Corporation President." *American Sociological Review* 46: 1-18.

Janis, I. 1972. *Victims of Groupthink*. Boston: Houghton-Mifflin.

Katz, R. 1981. "Managing Careers: The Influence of Job Longevity and Group Longevities." Pp. 154-181 in *Career Issues in Management*, edited by R. Katz. Englewood Cliffs, NJ: Prentice-Hall, Inc.

Keck, S. L. 1990. "Determinants and Consequences of Top Executive Team Structure." Unpublished doctoral dissertation, Columbia University, New York.

Kimberly, J. and M. Evanisko. 1981. "Organizational Innovation: The Influence of the Individual, Organizational, and Contextual Factors on Hospital Adoption of Technological and Administrative Innovations." *Academy of Management Journal* 24: 689-713.

Kotin, J. and M. Sharif. 1976. "Management Succession and Administrative Style." *Psychiatry* 30: 237-248.

Lawrence, P. and J. Lorsch. 1967. *Organization and its Environment*. Boston: Harvard University Press.

Lorsch, J. and S. Allen. 1973. *Managing Diversity and Interdependence*. Boston: Division of Research, Harvard Business School.

March, J. C. and J. G. March. 1977. "Almost Random Careers: The Wisconsin School Superintendency, 1940-1972." *Administrative Science Quarterly* 22: 377-409.

March, J. G. and H. Simon. 1958. *Organizations*. New York: Wiley.

Meyer, M. 1972. "Size and the Structure of Organizations: A Causal Analysis." *American Sociological Review* 37: 434-440.

Michel, J. and D. Hambrick. 1992. "Diversification Posture and the Characteristics of the Top Management Team." *Academy of Management Journal* 35: 9-37.

Miles, R. H. and K. Cameron. 1982. *Coffin Nails and Corporate Strategies*. Englewood Cliffs, NJ: Prentice-Hall.

Miles, R. and C. Snow. 1978. *Organizational Strategy, Structure, and Process*. New York: McGraw-Hill.

Miller, D. and J.M. Toulouse. 1986. "Strategy, Structure, CEO Personality, and Performance in Small Firms." *American Journal of Small Business* 10(3): 47-62.

Mintzberg, H. 1973. *The Nature of Managerial Work*. New York: Harper and Row.

———. 1979. *The Structuring of Organizations*. Englewood Cliffs, NJ: Prentice-Hall.

Mintzberg, H., D. Raisinghani, and A. Theoret. 1976. "The Structure of Unstructured Decision Processes." *Administrative Science Quarterly* 21: 246-275.

Murray, A. 1989. "Top Management Group Heterogeneity and Firm Performance." *Strategic Managment Journal* 10: 125-141.

Northcraft, G., T. Griffith, and C. Shalley. 1992. "Building Top Management Muscle in a Slow Growth Environment: How Different is Better at Greyhound Financial Corporation." *Academy of Management Executive* 6: 32-41.

O'Reilly, C., C. Caldwell, and D. Barnett. 1989. "Work Group Demography, Social Integration, and Turnover." *Administrative Science Quarterly* 34: 21-37.

Osborn, R. N., L. R. Jauch, T. N. Martin, and W. F. Glueck. 1981. "The Event of CEO Succession, Performance, and Environmental Conditions." *Academy of Management Journal* 24: 183-191.

Perrow, C. 1970. *Organizational Analysis.* Belmont, CA: Wadsworth.

Pfeffer, J. 1980. "A Partial Test of the Social Information Processing Model of Job Attitudes." *Human Relations* 33: 457-476.

_____. 1981. *Power in Organizations.* Boston: Pitman.

_____. 1983. "Organizational Demography." Pp. 299-357 in *Research in Organizational Behavior,* edited by L. Cummings and B. Staw. Greenwich, CT: JAI Press.

Pfeffer, J. and H. Leblebici. 1973. "Executive Recruitment and the Development of Interfirm Organizations. *Administrative Science Quarterly* 18: 449-461.

Pfeffer, J. and W. Moore. 1980. "Average Tenure of Academic Heads: The Effects of Paradigm, Size, and Departmental Demography." *Administrative Science Quarterly* 25: 387-406.

Pfeffer, J. and G. Salancik. 1978. *The External Control of Organizations.* New York: Harper and Row.

Pitts, R. A. 1977. "Strategies and Structures for Diversification." *Academy of Management Journal* 20: 197-208.

Porter, M. 1980. *Competitive Strategy.* New York: Free Press.

Pondy, L. 1969. "Effects of Size, Complexity, and Ownership on Administrative Intensity." *Administrative Science Quarterly* 14: 47-60.

Reinganum, M. 1985. "The Effect of Executive Succession on Stockholder wealth." *Administrative Science Quarterly* 30: 46-60.

Rogers, E. and F. Shoemaker. 1971. *Communication of Innovations.* New York: Free Press.

Rumelt, R. P. 1974. *Strategy, Structure, and Economic Performance.* Cambridge: Harvard University Press.

Salter, M. S. and W. A. Weinhold. 1978. "Diversification via Acquisition: Creating Value." *Harvard Business Review* 564: 166-176.

Shaw, M. E. 1976. *Group Dynamics: The Psychology of Small Group Behavior.* New York: McGraw Hill.

Smith, M. and M. C. White. 1987. "Strategy, CEO Specialization, and Succession." *Administrative Science Quarterly* 32: 263-280.

Snow, C. C. and L. G. Hrebiniak. 1980. "Strategy, Distinctive Competence, and Organizational Performance." *Administrative Science Quarterly* 25: 317-335.

Song, J. H. 1982. "Diversification Strategies and the Experience of Top Executives of Large Firms." *Strategic Management Journal* 2: 277-280.

Starbuck, W. 1976. "Organizations and their Environments." Pp. 1069-1123 in *Handbook of Industrial and Social Psychology,* edited by M. Dunnette. Chicago: Rand McNally.

Stevens, L., J. Beyer, and H. Trice. 1978. "Assessing Personal, Rate, and Organizational redictors of Managerial Commitment." *Academy of Management Journal* 21: 380-396.

Stinchcombe, A. S. 1965. "Organizations and Social Structure." Pp. 153-193 in *Handbook of Organizations,* edited by J. G. March. Chicago: Rand-McNally.

Taylor, R. 1975. "Age and Experience as Determinants of Managerial Information Processing and Decision Making Performance." *Academy of Management Journal* 18: 74-81.2

Terreberry, S. 1968. "The Evolution of Organization Environments." *Administrative Science Quarterly* 12: 590-613.

Thompson, J. 1967. *Organizations in Action.* New York: McGraw-Hill.

Tushman, M. L. and S. L. Keck. 1991. "Environmental and Organizational Context and Executive Team Characteristics: An Organization Learning Approach." Working Paper, Columbia University, Graduate School of Business, New York.

Tushman, M. and E. Romanelli. 1985. "Organizational Evolution: A Metamorphosis Model of Convergence and Reorientation." Pp. 171-222 in *Research in Organizational Behavior,* edited by L. Cummings and B. Staw. Greenwich, CT: JAI Press.

Tushman, M. L., B. Virany, and E. Romanelli. 1985. "Executive Succession, Strategy Reorientation, and Organization Evolution." *Technology in Society* 7: 297-314.

Von Glinow, M. A. and S. A. Mohrman. 1990. *Managing Complexity in High Technology Organizations.* Oxford University Press: Oxford.

Vroom, V. and B. Pahl. 1971. "Relationship Between Age and Risk-taking Among Managers." *Journal of Applied Psychology* 55: 399-405.

Wagner, W., J. Pfeffer, and C. O'Reilly. 1984. "Organizational Demography and Turnover in Top Management Groups." *Administrative Science Quarterly* 29: 74-92.

Wanous, J. and M. Youtz. 1986. "Solution Diversity and the Quality of Group Decisions." *Academy of Management Journal* 29: 49-58.

Wiersema, M. and K. Bantel. 1992. "Top Management Team Demography and Corporate Strategic Change." *Academy of Management Journal* 35: 91-121.

_____ . 1993. "Top Management Team Turnover as an Adaptation Mechanism: The Role of the Environment." *Strategic Management Journal* 14:485-504.

Williamson, O. 1963. "Managerial Discretion and Business Behavior." *American Economic Review* 53: 1032-1057.

Woodward, J. 1965. *Industrial Organization.* Oxford: Oxford University Press.

Zander, A. 1977. *Groups at Work.* San Francisco: Jossey-Bass, Inc.

Zenger, T. and B. Lawrence. 1989. "Organizational Demography: The Differential Effects of Age and Tenure Distributions on Technical Communication." *Academy of Management Journal* 322: 353-376.

SOCIAL DETERMINANTS OF CREATIVITY:
STATUS EXPECTATIONS AND THE EVALUATION OF
ORIGINAL PRODUCTS

Joseph Kasof

ABSTRACT

Creativity researchers generally have studied dispositional rather than situational causes of creativity. In particular, creativity researchers have largely ignored the subjective dimension of creativity—those factors that determine how an original product will be evaluated. This article examines one important social influence on the subjective dimension of creativity: how expectations based on status processes influence evaluations of original products. From status characteristics theory, hypotheses are derived concerning the relationship between evaluations of original products and the creator's diffuse status characteristics, specific status characteristics, and nonverbal behavior, as well as the status characteristics of third parties. Experimental evidence bearing on these hypotheses is reviewed.

Advances in Group Processes, Volume 12, pages 167-220.
Copyright © 1995 by JAI Press Inc.
All rights of reproduction in any form reserved.
ISBN: 1-55938-872-2

167

The history of creativity research is the tale of a quest for dispositional rather than situational causes. From its beginning, creativity research was conducted primarily not by sociologists, economists, anthropologists, or social psychologists, but by personality and cognitive psychologists searching for the distinctive traits of "creative people." Only very recently have creativity researchers begun searching the social environment for causal influences (e.g., Amabile 1983; Simonton 1984), and even these efforts generally have ignored distinctively sociological influences.[1] The field's skewed emphasis on creators' dispositions was revealed in a recent bibliographic study by Wehner, Csikszentmihalyi, and Magyari-Beck (1991). After carefully selecting database search terms relevant to creativity in various areas of specialization, Wehner and colleagues searched the 1986 *Dissertation Abstracts Online* for dissertations on creativity. A sample of the dissertations was then categorized in terms of the aspect of creativity that was studied (trait, process, or product) and the level at which it was studied (individual, group, organization, or culture). Results showed that the dissertations were conducted more commonly at the individual level than at any other level. Summarizing their findings, Wehner, Csikszentmihalyi, and Magyari-Beck concluded:

> ...It is clear that certain important dimensions of creativity do not seem to be studied by doctoral students at all. There are almost no dissertations written on the traits or products of creative cultures, and on the traits or products of creative groups. The almost total absence of studies of creative groups is particularly strange, especially when one considers the theoretical importance of brainstorming procedures or the historical importance of creative groups like the Cavendish laboratory in physics or the Impressionists' coterie in art (p. 270).

Moreover, Wehner, Csikszentmihalyi, and Magyari-Beck almost certainly *under*estimated the field's emphasis on dispositional causes by oversampling the less frequently occurring disciplines (e.g., business and economics, political science) and undersampling the more dominant disciplines (e.g., psychology), which were the most heavily concentrated at the individual level. In psychology, the field in which efforts to understand creativity have been most highly concentrated, fully 87 percent of the creativity dissertations were focused on the individual level; not a single one focused on groups or culture.

In contrast to the dispositional emphasis of past creativity research, this paper examines creativity from a sociological perspective. Creativity researchers generally define the creative product as a product that satisfies two basic criteria: (1) the creative product must be unusually original, rare, novel, statistically infrequent and (2) it must be approved, accepted, valued, considered "appropriate" or "good." By this definition, creativity is not a purely objective, fixed property of the creative product that holds always and everywhere; in part, it involves a subjective judgment that is conferred on the original product (Amabile 1982; Csikszentmihalyi 1988, 1990; Csikszentmihalyi and Robinson 1986; Gardner 1993; Simonton 1988; Weisberg 1986). The first criterion, which might be called the *originality* requirement, is objective; the second criterion, the *evaluation* requirement, is subjective. Products that are approved but unoriginal, original but unapproved, or unoriginal and unapproved fail to meet both the objective and subjective requirements and therefore are by definition not creative. Among the products that do meet both criteria, the *degree* of creativity depends jointly on the degree of originality and the degree of approval the product receives.

Yet although creativity, like beauty, is located partly in the "eye of the beholder," dispositionally oriented creativity researchers have generally regarded variation in the evaluation of original products not as something to be explained, but as something to be avoided (for exceptions see Csikszentmihalyi 1988, 1990; Getzels and Csikszentmihalyi 1976, chap. 7; Runco and Charles 1993; Runco and Smith 1992; Runco and Vega 1990; Simonton 1984, chap. 7). To this end researchers have developed "objective creativity tests" for which inter-rater reliability is nearly perfect and have used "blind" judges to evaluate creations in ignorance of the creator's race, sex, physical attractiveness, and other "contaminating" information that might influence evaluation. Although this approach has important merits, in stripping the laboratory of powerful social influences it reduces ecological validity, creates a degree of inter-rater reliability vastly surpassing that of the real world, obscures the subjective dimension of creativity, and permits creativity research to proceed harmoniously without attending to the messy business of historical, cross-cultural, and individual variation in the evaluation of original products. Consequently, whereas creativity researchers have learned much about the traits of successful creators and about some of the situational factors that affect the production of creative works, relatively little is known about the situational factors that influence the *reception* of original products—for example, social factors that influence the normative criteria by which persons evaluate original products (including variation in such criteria across individuals, groups, cultures, and history), and systematic departures from such normative criteria (i.e., biases) in the evaluation of original products. This gap in knowledge is important not only because the reception of an original product critically influences whether and to what extent the product is creative, but also because an original product's reception can facilitate or inhibit the future production of creative works by the creator and by other potential creators (Kasof in press).

Status is an important but overlooked variable that affects both the objective and subjective dimensions of creativity. Opportunities for creation are constrained profoundly by the creator's status. Status differences between the sexes, for example, have led to a great imbalance between the number of male creators and female creators throughout history (e.g., Bare and Ptacek 1988; Simonton 1992b; Tuchman and Fortin 1984). Slavery, the lowest of all statuses, nearly always involves deprivation in opportunities for originality (Patterson 1982). More subtly, socioeconomic status affects workers' valuation of conformity versus originality in themselves and their children (Kohn, Naoi, Schoenbach, Schooler, and Slomczynski 1990; Kohn and Schooler 1969, 1982).

This article is concerned primarily with how status affects the subjective dimension of creativity—that is, how status affects evaluations of the creator's work. While status-based judgment is a pervasive fact of everyday life and has been extensively studied by social psychologists, researchers have seldom considered it in relation to creativity. In this paper I will argue that status processes affect the reception of original products and thereby partly determine what is creative and what is not, and how creative an original product becomes. In the first section of this paper, I briefly summarize status characteristics theory and research. In the second section, I develop hypotheses linking status-based expectations to the evaluation of original products. In the third section, I review experimental evidence bearing on these hypotheses.

STATUS CHARACTERISTICS THEORY

Joseph Berger and his colleagues have formulated and tested the theory of status characteristics and expectation states to explain how status affects discrimination in task-oriented groups (Berger, Fisek, Norman, and Zelditch 1977; Berger and Zelditch 1985). The theory proposes that persons who are working on a group task and who are motivated to succeed use information about a group member's status in deciding whether to accept or reject the member's contributions. This decision is influenced by performance expectations, which are based on the group members' status characteristics. Other things being equal, group members who have high status characteristics will receive and will take more opportunities to contribute to the group, and whatever contributions they make will be valued more highly than those made by group members whose status characteristics are low.

Status characteristics theory further proposes that status characteristics vary in generality. *Diffuse* status characteristics are social characteristics that involve at least two differentially valued states that are associated with expectation states for performance on a relatively wide variety of tasks. Examples of diffuse status characteristics include race (Howard and Pike 1986; Sev'er 1989), gender (Ridgeway 1992), physical attractiveness (Eagly, Ashmore, Makhijani, and Longo 1991; Kalick 1988; Webster and Driskell 1983), age (Baker 1985, 1988;

Boyd and Dowd 1988), educational attainment (Markovsky, Smith, and Berger 1984), and socioeconomic status (Howard and Pike 1986). Each of these characteristics has states that are generally regarded as positive or negative and for which there exist widespread performance expectations related to a large variety of tasks.

Specific status characteristics also are social characteristics that involve at least two differentially valued states, but they are associated with expectation states for performance on a narrower range of tasks. Examples of specific status characteristics include one's ability at specific tasks such as physics, poetry, and basketball. Nobel-laureate physicists, Pulitzer Prize-winning poets, and professional basketball players have higher specific status characteristics in their respective domains than do nonphysicists, bad poets, and basketball players cut from the team. The performance expectations associated with specific status characteristics affect discrimination only in the relatively narrow range of tasks to which the expectation states refer.

Status characteristics theory has several basic assumptions and scope conditions. First, unless a status characteristic is understood to be irrelevant to a given task, status characteristics that differentiate individuals will become salient whereas status characteristics that are shared by all group members will not become salient and thus will not result in discrimination. For example, on tasks in which race is thought to be relevant, race should be salient and influential in mixed-race interactions but not in same-race interactions. Status characteristics theory further assumes that group members will proceed as if salient status characteristics are relevant to the performance of a task unless the status characteristic is already believed to be irrelevant or unless its relevance is explicitly denied; in other words, members tend to behave as if the "burden of proof" is on those who wish to demonstrate a status characteristic's irrelevance rather than on those who wish to demonstrate its relevance to the performance of a given task. Multiple status characteristics, moreover, are processed not by ignoring inconsistent characteristics but by combining them into aggregate expectation states; for example, general performance expectations for Black attorneys should be higher than for Black custodians but lower than for white attorneys. Finally, when the salient status characteristics that are considered task-relevant are aggregated for self and for other, the emerging differences in power and prestige are a direct function of the actor's expectation (dis)advantage relative to the other. As it is currently formulated, status characteristics theory applies only to collective task-oriented situations involving interaction between actors who value successful completion of the task. The actors may be either "interactants" (actors engaged directly in social interaction) or "referent actors" (actors not engaged directly in social interaction but whose status characteristics influence interactants' expectations for each other).

Status characteristics theory has stimulated a good deal of experimental research, most of which has been supportive (see Berger and Zelditch 1985; Webster and Foschi 1988). For example, group members who have positive diffuse status characteristics enjoy an "interaction advantage" (e.g., Ridgeway 1982; Webster and Driskell 1978); members who have positive specific status characteristics are similarly advantaged (e.g., Wagner and Berger 1982; Webster 1977); and the expectation disadvantage resulting from low diffuse status characteristics can be reversed by presentation of high specific status characteristics (e.g., Cohen and Roper 1972; Ridgeway 1982).

STATUS CHARACTERISTICS AND EXPECTATIONS OF CREATIVE BEHAVIOR

The scope conditions of status characteristics theory encompass many situations in which original products are evaluated. When interactants are engaged in a collective task of judging an original product, and when the creator is either an interactant or a referent actor, the evaluation of the original product may be influenced by the creator's status characteristics as well as by the status characteristics of interacting or referent noncreators.

The evaluation of original products is a task that often takes place in collective task situations in which creators interact. When scientists or musicians work together on collaborative projects, for example, and when people work in brainstorming groups or product-innovation groups, the members of the group are interacting for the purpose of producing a collective creative product and, covertly if not overtly, they are evaluating the contributions of the group members.[2] In such situations, the status characteristics of the group members theoretically should influence the evaluation of the members' original contributions.

Additionally, the task of evaluating an original product often takes place in situations in which the creator is a referent actor and noncreating interactants are engaged in a collective task of evaluating the creator's work. This occurs, for example, when faculty review committees discuss the work of job applicants or of faculty members being considered for promotion, when two people at a cafe discuss a film, when judges meet to discuss creations that have been entered into a contest, when record company executives, talent scouts, and managers discuss whether to sign a musical creator to a label, and when docents try to persuade gallery- or museum-goers of the quality of the displayed art. In these situations, the evaluation of the referent creator's original products theoretically should be influenced both by the referent creator's status characteristics and by the status characteristics of the noncreating interactants.

In a faculty review committee, for example, a high-status committee member's opinion of a referent creator's work should generally be more influential than a low-status committee member's opinion.

In addition to these situations that fall clearly within the scope of status characteristics theory, the scope of status characteristics theory may be extended to encompass many other creativity-judgment situations. In its current formulation, status characteristics theory defines "interaction" as social behavior that involves potentially mutual influence of the same kind between two actors and the tasks to which the theory applies are only those that are "collective," that is, tasks in which interactants share the same task-goal. At present, in other words, status characteristics theory is limited to judgment-tasks that are both interactive and collective. There are, however, two kinds of *non-interactive judgment-tasks* for which the basic logic of status characteristics theory should also apply in many situations. First, *non-interactive collective judgment* tasks are those situations in which two or more actors judge a performance output but do so in a manner that does not involve bidirectional influence of the same kind. For example, when an editor at a scientific journal or publishing house evaluates submitted manuscripts by weighing the opinions of various reviewers to whom the manuscripts were sent, or when a gatekeeper at a funding agency judges grant proposals by soliciting the opinions of reviewers, multiple actors are engaged in a collective task of evaluating the creator's original product, but the actors' judgments are not mutually influential; in making their judgments, editors weigh the opinions of reviewers, but reviewers' judgments are not made by weighing the opinions of editors. Second, *non-interactive noncollective judgment tasks* are those situations in which persons acting noncollectively judge a performance output and do so in a manner that does not involve bidirectional influence of the same kind. For example, when a lone judge chooses among multiple creations that have been entered into a contest, or when an individual's evaluation of a film or painting is influenced by a critic's review in a newspaper, or when an editor at a scientific journal evaluates submitted manuscripts without soliciting reviews from others, the judgment task is neither interactive nor collective.

To encompass such non-interactive judgment-tasks, I would like to propose an extension of the scope conditions of status characteristics theory. All else being equal, when at least two actors, o^1 and o^2, make performance outputs that are evaluated by one non-interacting actor, p, whose task-goal is to make unit evaluations of those performance outputs, then the higher the state of salient status characteristics of o^1 relative to o^2, the higher p's performance expectations will be for o^1 than o^2. The higher p's performance expectations for o^1 than for o^2, the more positively p will evaluate future performance outputs of o^1 than o^2, the more influence p will accept from o^1 than from o^2, and the more future opportunities for performance output p will grant o^1 relative to o^2. Thus, for example, non-interacting judges should generally evaluate paintings more positively when they are attributed to a high-status artist than a low-status artist. Moreover, just as previous formulations of status characteristics theory predict that interactants may be influenced by referent actors' status characteristics, so the proposed scope-extension holds that non-interacting actors may be influenced by referent actors' status characteristics. The higher the status characteristics of the referent actor, the greater should be the referent actor's influence on p's evaluation of o^1 and o^2. For example, an editor's evaluation of a submitted manuscript should be influenced more strongly by the opinion of a high-status reviewer than by the opinion of a low-status reviewer.[3]

Within the proposed expanded scope conditions of status characteristics theory, the potential for judgments of original products to be influenced by status characteristics is high. Let us now consider in greater depth how expectations of creative behavior are related to status characteristics.

Diffuse Status Characteristics and Expectations of Creative Behavior

Among the diffuse status characteristics that evoke stereotypic expectations related to creative behavior are physical attractiveness, age, social class, gender, race, sexual orientation, and nationality. The relevance of these stereotypes to creativity is either direct, when stereotypes have been measured by asking subjects to rate groups in terms of traits such as *creative* or *innovative*, or indirect, when stereotypes have been measured by asking subjects to rate groups in terms of traits that are widely believed to be related to creativity, such as *flexibility, open-mindedness, inquisitiveness,* and *risk-taking* (Katz and Giacommelli 1982; Rossman and Gollub 1975; Runco and Bahleda 1986; Sternberg 1985). Physically attractive persons are stereotypically thought to be more creative, more flexible, more curious, and generally more intellectually competent than are physically unattractive people (Clifford and Hatfield 1973; Eagly, Ashmore, Makhijani, and Longo 1991; Feingold 1992; Miller 1970). Elderly persons are widely considered to be low in innovativeness, flexibility, open-mindedness, risk-taking, and general intellectual competence (Crockett and Hummert 1987; Kite and Johnson 1988; Rosen and Jerdee 1976). Working-class individuals are generally expected to be less creative than are professionals (Feldman 1972). Males are commonly expected to be more intelligent, scientific, and mathematical than females, but less artistic and literary (e.g., Fernberger 1948; Ward and Balswick 1978), although research has not yet determined whether these stereotypes pertain to creativity-relevant behavior versus other behavior within these domains. Homosexual men are expected to behave more creatively, especially in the arts, than heterosexual men (Lord, Lepper, and Mackie 1984; Simon, Glassner-Bayerl, and Stratenwerth 1991). Whites are widely expected to be generally more intellectually competent but less musical than Blacks (Feldman 1972; Karlins, Coffman, and Walters 1969). Several national stereotypes are related to creativity: Indians expect the Chinese to be highly artistic (Sinha and Upadhyaya 1960), Iranians expect Americans to be highly musical (Beattie, Agahi, and Spencer 1982), and Americans seem to view the Japanese as being technically competent but low in creativity.

Because diffuse status characteristics affect expectations of creative behavior, creators who have high diffuse status characteristics should enjoy an expectation advantage over creators who have low diffuse status characteristics. This expectation advantage should cause observers to be favorably biased in evaluating the work of creators who have high diffuse status characteristics relative to those who have low diffuse status characteristics. For example, all else being equal, in experimental brainstorming groups and product-innovation groups, original ideas should be evaluated more positively when attributed to physically attractive creators than to physically unattractive creators, more positively when attributed to male creators than to female creators (at least in the domains of science and mathematics and in gender-neutral tasks), and more positively when attributed to high-SES creators than to low-SES creators.

Specific Status Characteristics and Expectations of Creative Behavior

Many status characteristics are linked specifically to expectations of creative behavior. Honorific distinctions such as the Pulitzer and the Nobel Prize, the "Genius" awards of the MacArthur Foundation, and membership in prestigious groups such as the National Academy of Science are specific status cues that increase the expectation of domain-specific creative behavior. The distinguished creator's specific status characteristics should heighten others' expectations of creative behavior and should bias them favorably in evaluating the creator's work. All else being equal, therefore, in brainstorming groups, in teams of collaborative creators, and in other judgment situations, an original product should be evaluated more positively when it is attributed to a creator who is considered highly creative than when attributed to a creator who is thought to be less creatively talented.

Consider, for example, the experiments conducted by Sachdev and Bourhis (1987, 1991). In a group setting, undergraduates worked individually on a standardized test purportedly measuring creative ability. While the researcher was supposedly scoring these tests, students worked on a second creativity task, in which they individually created titles for an abstract painting. On a random

basis, students were then given bogus positive or negative feedback on the firs
creativity test and subsequently were asked to evaluate painting titles attributec
to two anonymous subjects who had allegedly scored high and low on the initia
test of creative ability. The proposed scope-extension of status characteristic:
theory predicts that students would evaluate the titles more positively wher
attributed to students high in creative ability than when attributed to student:
low in creative ability. This in fact was the result, both for students who had
received negative feedback and for students who had received positive feedback
on the initial creativity test.

Forgery

The impact of specific status-based expectations on evaluations of origina
products is illustrated in the history of art forgery. For example, Fine (1983)
described the case of an obscure Dutchman named Han van Meegeren, whc
stunned the art world in the late 1930s by presenting for sale *Christ at Emmaus,*
a previously unknown painting which he falsely attributed to the seventeenth-
century master Vermeer. Van Meegeren claimed to be serving as broker for
a prominent unnamed Italian family in whose possession the painting had
remained for generations. *Christ at Emmaus* quickly won high acclaim from
the arts community: The foremost Dutch expert on seventeenth-century art,
Abraham Bredius, to whom the painting was submitted for authentication,
enthusiastically declared it a work of genius, indeed Vermeer's finest
masterpiece, and eventually it was purchased by the prestigious Boymans
Museum in Rotterdam. Soon, van Meegeren produced other previously
unknown "Vermeers" as well as other paintings that he falsely attributed to
Peter de Hooch. These paintings too received accolades from art critics.

Only years later was it discovered that these paintings were actually created
by van Meegeren himself. A struggling artist whose work, when it was
attributed to him, had been rejected and ignored by the critics, van Meegeren
had executed the hoax to show the world that the critics were biased by specific
status characteristics—that the falsified signature of an established master
could elevate an undistinguished artist's work to the level of world-class
masterpiece. After van Meegeren's confession, his "masterpieces" were
promptly removed from the hallowed walls they had formerly distinguished.

This and other art forgeries suggest the importance of specific status-based
expectations in evaluations of original products. Approval of van Meegeren's
art rose and fell not with any objective characteristics of his paintings but
rather with the presumed status of their creator. Decades after van
Meegeren's forgeries were removed from prestigious collections, authentic
Vermeers remain fixed to their walls, still benefiting from the bias that once

had briefly transformed the unknown works of a rejected artist into products of genius.

Institutional Status

Although it has not yet been investigated as such, institutional status seems to often function as a "status cue" (Berger, Webster, Ridgeway, and Rosenholtz 1986) for specific status characteristics of creators. For example, a doctorate from a highly ranked university bestows an expectation advantage on a scientist (Long, Allison, and McGinnis 1979). Similarly, academic scientists affiliated with top-ranked institutions enjoy an expectation advantage over scientists affiliated with lower-ranked institutions. This fact partly explains why mobility from a department of lower prestige to one of higher standing increases the number of citations given to such scientists' previously published work (Allison and Long 1990), why scientists affiliated with highly ranked universities produce work that is regarded more positively than that of scientists affiliated with less prestigious universities (Bakanic, McPhail, and Simon 1987; Cole and Cole 1973; but see Cole, Rubin, and Cole 1979), and why eminent scientists are generally affiliated with highly prestigious research institutions (Crane 1965; Helmreich, Spence, Beane, Lucker, and Matthews 1980; Simonton 1992a; Zuckerman 1977). Low institutional status explains the delay in the recognition of many important creations by scientists affiliated with institutions of low standing or by scientists unaffiliated with any institution; examples include the works of Mendel, Ohm, and Abel (Barber 1961, p. 600).

In art worlds, too, institutional prestige enhances domain-specific status. Artistic reputation is enhanced by graduating from a prestigious art school, by being represented in prestigious galleries and museums, and by belonging to a prestigious art faculty (Getzels and Csikszentmihalyi 1976; Greenfeld 1989; Maquet 1986). As in science, affiliation with a low-ranked institution or with no institution at all may bias negatively the application of normative aesthetic criteria, thereby preventing or delaying recognition of artistic works that otherwise might be positively accepted. The reception of Van Gogh's art is one likely example.

Category Status

Aside from the creator's status characteristics, original products themselves belong to categories that vary in status. Thus many such categories may function as specific status cues (Berger et al. 1986) which lead to differing performance expectations and thereby influence evaluations of the original products. "High cultural" forms, for example, have generally higher status than do "popular cultural" forms (DiMaggio 1987; DiMaggio and Useem 1978; Gans 1974). In contemporary Western cultures, ballet, opera, classical music, and novels are more prestigious categories of creation than are tap dancing, graffiti or comic art, country-and-western music, and children's stories.

Within the major categories of original products, subcategories also vary in status. Within the domain of science, for example, physics has higher standing than does sociology. Even within scientific fields, subfields and areas of specialization vary in status. Within sociology, for example, research on social stratification has higher standing than does ethnomethodology or the sociology of knowledge, and these subdisciplinary differences in status may lead to corresponding differences in performance expectations and, thereby, biases in evaluation (cf. Bakanic, McPhail, and Simon 1987).

Personality Attributes

Stereotype-based expectations pertaining specifically to creative performance may arise from the creator's personality characteristics (cf. Johnston 1985), as in stereotypic conceptions of the "creative person." As noted earlier, certain cognitive and personality traits are stereotypically associated with creative talent. For example, high risk takers are expected to behave more creatively than low risk takers (Jellison and Riskind 1970; Jellison, Riskind, and Broll 1972), and greater creativity is expected from individuals who are unconventional, independent-minded, inquisitive, humorous, intellectual, rebellious, and intuitive, than from people who lack these traits (Katz and Giacommelli 1982; Rossman and Gollub 1975; Runco and Bahleda 1986; Sternberg 1985). Thus, even in the absence of information about a creator's domain-specific status characteristics, a "creative person" schema or stereotype, activated when a person comes close to fulfilling the prototype, may cause observers to engage in biased top-down information processing in which they "see people as being creative according to how they act or present themselves and not according to what they produce" (Katz and Giacommelli 1982, p. 20).

Third-Party Status Characteristics and Expectations of Creative Behavior

An actor's expectations of creative behavior and evaluation of an original product may be influenced not only by the status characteristics of the creator, but also by the status characteristics of persons other than the creator. These *third parties* can be either interactants in collective judgment-tasks (e.g., when tenure review committees composed of members differing in status discuss the work of a faculty member) or actors in non-interactive judgment-tasks (e.g., when a lone actor is influenced by critics' reviews).

History offers many examples of third parties' status characteristics facilitating or inhibiting actors' approval for original products. According to Kuhn (1957), the acceptance of Copernicanism was delayed substantially by the opposition of the astronomer Brahe, in part because Brahe's prestige made him highly influential. Approval for Bell's telephone apparently was facilitated by the high status of the Emperor of Brazil, who had visited Bell's class for the deaf at Boston University and whose subsequent recognition of Bell at the Centennial Exposition in Philadelphia influenced the judges in favor of the then-obscure inventor (Madigan and Elwood 1983, p. 206). Approval for Einstein's early work and for Darwin's theory of evolution were similarly facilitated by the endorsement of eminent scientists (Cohen 1985).

Collaboration

Collaboration with high-status creators may function as an implicit third-party endorsement and status cue (Berger et al. 1986) and thereby bias judges favorably in evaluating the creator's work. For example, graduate students with whom highly regarded faculty have chosen to work may benefit from an expectation advantage and become more likely to gain approval for their own work. In support of this hypothesis, the prestige of a graduate student's mentor is related directly to the prestige of the graduate student's first faculty position, even when graduate student productivity and departmental status are held constant (Allison and Long 1987; Long, Allison, and McGinnis 1979; also see Zuckerman 1977).

Gatekeeping

The status characteristics of "gatekeepers" is another way in which a third party's status characteristics can affect actors' expectations and thereby their evaluations of original products. The impact of a critic's published assessment of an original product on others' evaluations of the original product increases with the critic's status (see Crane 1967; Shrum 1991); indeed, merely being reviewed by a high-status critic or in a high-status publication confers a certain prestige on the original product and its creator, even if the review is not positive. Better artwork is expected from artists whose work has been purchased by celebrities and respected collectors (Getzels and Csikszentmihalyi 1976; Maquet 1986). Scholarly articles published in prestigious journals have an expectation advantage over papers published in poorly regarded journals (Zuckerman and Merton 1971). Paintings hung in prestigious museums and galleries such as the Louvre or the Museum of Modern Art should enjoy similar advantage. Cuisine served at prestigious restaurants, plays performed in prestigious theaters, poetry published by respected presses or in prestigious journals, interior designs featured in prestigious buildings—each should tend to be evaluated with a favorable bias.

Performance

The status characteristics of noncreators performing the creator's work also may function as a status cue and exert a third-party status effect. When a high-status performer performs someone else's creations, the creator's reputation should benefit from the association, and the creator's work should be evaluated with a favorable bias. In performing works created by others, high-status singers enhance a songwriter's reputation, prestigious orchestras enhance a composer's standing, famous comedians enhance a comedy writer's reputation, and prestigious dance companies enhance a choreographer's status. Once the creator's status is so enhanced, others should be biased favorably in evaluating the creator's work.[4]

The Cultural and Historical Relativity of Status Characteristics

Although the literature on status characteristics has been concerned primarily with contemporary American society, many status characteristics are historically and culturally bound. In contrast to our culture, many premodern societies accord high status to old age (Palmore and Manton 1974). The Indian caste system and medieval European feudalism determined individuals' diffuse status characteristics. Among the Mundugumor of New Guinea, boys born with the umbilical cord wrapped around their neck are expected to become artists; their paintings are approved, while paintings made by others are ignored or rejected (Mead 1935). During the Middle Ages and the Renaissance, artistic status was influenced strongly by membership and position in guilds (Getzels and Csikszentmihalyi 1976, p. 186). Further, the status of different categories of creative endeavor varies across cultures and history. The novel as a literary form increased in status during the nineteenth century (Tuchman and Fortin 1984; Watt 1957), whereas etching declined in status (Lang and Lang 1988). Each of these historically and culturally specific bases of status defines the particular hypotheses relating status-based expectations to judges' evaluations.

Nonverbal Status Cues and Expectations of Creative Behavior

Various nonverbal and paralinguistic behaviors are associated with status characteristics. Rapid speech connotes greater intellectual competence than does slow speech (Miller, Maruyama, Beaber, and Valone 1976). Group members expressing unconventional ideas are more influential when they deliberately take the head seat at a table than when they take a side seat or are merely assigned to sit at the head of the table (Nemeth and Wachtler 1974). Status also is linked with eye gaze (Ridgeway, Berger, and Smith 1985; Rosa and Mazur 1979), verbal response latency (Mazur, Rosa, Faupel, Heller, Leen, and Thurman 1980; Ridgeway et al. 1985; Rosa and Mazur 1979; Willard and Strotdbeck 1972), and conversational interruption (Kollock, Blumstein, and Schwartz 1985; Leffler, Gillespie, and Conaty 1982).

Cecilia Ridgeway and Joseph Berger recently incorporated such nonverbal behaviors into status characteristics theory (Berger et al. 1986; Ridgeway and Berger 1986; Ridgeway et al. 1985). They proposed that individuals who maintain an expectation advantage behave in a relatively confident and assertive manner, whereas those who maintain an expectation disadvantage behave less confidently. Thus behaviors expressing high task confidence are "high task cues" that observers associate with high status and superior task performance, whereas behaviors expressing low task confidence are "low task cues" that observers associate with low status and inferior task performance. Thus when actors speak confidently, observers generally assume that such people know what they are talking about. Conversely, when actors speak slowly and with averted gaze, concede the floor readily when interrupted, and otherwise behave without confidence, observers infer low task ability. Experimental research supports these and related hypotheses (e.g., Berger et al. 1986; Dovidio, Brown, Heltman, Ellyson, and Keating 1988; Dovidio, Ellyson, Keating, Heltman, and Brown 1988; Leffler et al. 1982; Mohr 1986; Ridgeway 1987; Ridgeway et al. 1985; Ridgeway and Diekema 1989; Sev'er 1989; Wood and Karten 1986).

If task cues affect performance expectations, and if performance expectations affect evaluations of original products, then task cues should affect evaluations of original products. Creators who bristle with confidence should enjoy an expectation advantage over creators who appear meek and uncertain. Creators who speak rapidly and without hesitation, for instance, should win greater approval for their ideas than should creators who speak slowly, quietly, and with frequent hesitations and hedges. In product-innovation groups, members' evaluations of one another's contributions should be affected by task cues such as verbal response latency, speech rate, and eye gaze.

The expectation advantage that is conferred upon confident creators predicts the finding from dispositional research that creators whose work is highly regarded are highly confident. Creators whom others judge to be creative are unusually likely to describe themselves as *dominant, egotistical, self-confident,* and *snobbish* and to score high on other measures of self-confidence, dominance, narcissism, and self-assertiveness (e.g., Cattell and Drevdahl 1955; Domino 1970; Gough 1961, 1979; Harrington 1975; Helson 1971, 1980, 1985; Kaduson and Schaefer 1991; MacKinnon 1964).

Individual differences in demeanor should affect posthumous variation in the evaluation of original products. Many creators' work was widely rejected or ignored by their contemporaries, only to become famous after their deaths (e.g., Barber 1961). In many such cases the creators themselves may have inadvertently reduced contemporaneous approval for their creations by instilling negative performance expectations in others with their averted gaze, slow speech, informal or unkempt appearance, and other low task cues (Kasof in press). Such task cues, however, influence the creator's contemporaries more strongly than they influence posthumous evaluators. Decades after Van Gogh's death, for example, museum-goers and critics were unaffected by the artist's unconfident demeanor, self-belittlement, and tattered clothing, but these low task cues apparently influenced his contemporaries' expectations, causing him to be regarded as "somebody who has no position in society and never will" (Wallace 1969, p. 33). After Van Gogh's death, however, vivid memories of the unconfident derelict faded, and with them faded his expectation disadvantage. Eventually his paintings could be viewed with less unfavorable prejudice; only then was Van Gogh "discovered."

Conversely, many creations have received great acclaim during the creator's lifetime, only to be forgotten, rejected, or approved less enthusiastically by subsequent generations. In many such cases, creators may have instilled positive expectations in others by their rapid, unhesitating speech, their direct eye gaze, their formal attire, and other high task cues. Because task cues influence a creator's contemporaries more strongly than they influence posthumous evaluators, the expectation advantage caused by the confident creator's high task cues should fade after the creator's death. Insofar as approval for a creation stems from the creator's demeanor, the renown of the creation should pass with the years following the creator's death.

Low task cues also should affect delayed recognition within the creator's lifetime. Consider, for example, Einstein's important early work, which remained obscure for several years after its publication (Cohen 1985; Hoffmann 1972). Because Einstein published these early papers when he was an unknown clerk without a doctorate or a university affiliation, physicists reading his publications in 1905 could not have had very high status-based expectations. Yet even after Einstein completed his dissertation and received a faculty position, his nonverbal self-presentation may have further delayed recognition of his work: As a young instructor "he made no attempt to improve his appearance or alter his manner to accord with academic custom" (Hoffmann 1972, p. 87), he wore "inappropriately casual clothes" (p. 150), and his appearance was so unkempt that a former teacher mistook him for a beggar (Ochse 1990, p. 90). Years later, after he became extremely famous, Einstein's retiring and self-deprecating manner, his preference for casual attire and neglect of grooming, his slow and high-pitched speech, and his generally unheroic demeanor may have endeared him to his admirers and contributed to his legend; but in his early years at the patent office and as a lecturer at Bern, these very qualities probably postponed his stardom. The expectation disadvantage he generated is illustrated in this anecdote:

> Among the students in Bern in those days were many Russian Jews, poor, ill-clad, unkempt, and therefore frowned upon. Einstein's sister, Maja, tells of an incident that suggests what sort of outward impression Einstein made on the authorities. She was a student at Bern University at the time. Wishing to hear one of Einstein's lectures, she asked the doorman which room her brother, Dr. Einstein, was lecturing in. Looking at the neat young lady before him, the doorman blurted out in utter astonishment, 'What? That...Russ is your brother?' (Hoffmann 1972, pp. 87-88).

Differences in task cues also should affect geographical variation in the evaluation of original products. Some creations receive greater approval in the creator's locale than elsewhere; others receive greater approval elsewhere than in the creator's locale. Such geographical variation may result from geographical variation in performance expectations resulting from the creator's task cues. For example, publications by scientists who lack confidence may be regarded skeptically by departmental colleagues but received more enthusiastically by distant scientists who do not know the creator personally. Conversely, publications by scientists who behave very confidently should be evaluated more enthusiastically by departmental colleagues and personal acquaintances than by others who do not know the dynamic creator. Insofar as approval for a creation stems from the creator's task cues, that approval should decline with the distance between creator and judge.

EXPERIMENTAL EVIDENCE

Although status characteristics theory has been tested extensively, it has not yet been tested as such in experiments in which the evaluation of original products was a dependent variable. Nevertheless, dozens of experiments on brainstorming, discrimination, prestige suggestion, and teacher's expectancies have been conducted in which the dependent variable included the evaluation of original products and in which the independent variable included the alleged creator's status characteristics. In some of these studies, the evaluation of original products was a collective task involving interacting judges; such studies satisfy the scope conditions of status characteristics theory. In the majority of studies, however, the evaluation of original products was made by non-interacting actors; these studies satisfy the extended scope conditions of status characteristics theory proposed earlier. What follows is a brief summary and critique of this experimental literature.

To locate experiments for this review, I used the following search procedure. Diffuse status characteristics included gender, physical attractiveness, race, ethnicity, age, and socioeconomic status (including its components, such as educational attainment). Specific status characteristics were considered to be knowledge or creative accomplishment in the domain to which the original product in the experiment belongs. Original products were defined as novel responses to heuristic rather than algorithmic tasks (cf. Amabile 1983); this definition encompasses studies in which subjects evaluated drawings, essays, musical compositions, lithographs, poems, paintings, pottery, prose passages, stories, and ideas generated in brainstorming sessions, but excludes studies in which subjects evaluated leadership behavior, job applications, memos, perceptual judgments, resumes, and so forth. Relevant studies were then sought by (a) conducting an online literature search of *PsycLIT* and *Sociofile* using appropriate combinations of numerous database search terms, such as *status, age, gender, prestige, stereotype, group,* and *discrimination*; (b) using the references of relevant articles to find previously published studies; (c) using *Social Science Citation Index* to find later studies that cited a relevant study; and (d) manually searching major sociology, social psychology, and creativity journals. The literature search was limited to studies published before 1991.

Interacting Judges' Evaluations of Original Products

Although much research has been done on group creativity (e.g., Mullen, Johnson, and Salas 1991), very little research has been done on social processes affecting the evaluation of ideas generated in such groups. The existing research, though extremely limited, supports status characteristics theory. Collaros and Anderson (1969) found that when members of brainstorming groups were misled to believe that other members of their group were higher than themselves in a specific status characteristic (prior experience in brainstorming groups), they evaluated their own ideas less positively, felt more inhibited to contribute ideas to the group, and in fact contributed fewer ideas. Torrance (1963) found that after elementary-school students engaged in group brainstorming to generate creative uses for male-typed toys, the group members evaluated boys' contributions more positively than girls' contributions even when there was no sex difference in the number of ideas contributed; because the ideas were not rated by blind judges, however, it remains possible that the boys' contributions were higher in quality though not in quantity. In a study of mixed-sex brainstorming groups working on an apparently gender-neutral task, Ruback, Dabbs, and Hopper (1984) found that although male and female undergraduates did not differ in the number or length of their vocalized contributions, females were perceived by their peers as having talked less and as having demonstrated less "leadership," which was operationalized by a scale including such questions as "How good were this person's ideas?" and "How much did this person contribute to the group?"

Several other studies have been conducted with interacting creators in which explicit evaluations per se were not measured but in which other behaviors predicted by status characteristics theory were measured. Several experiments have found that status in brainstorming groups varies directly with the number of ideas the members contribute and with the frequency and length of the members' vocal output (Jablin and Sussman 1978; Jablin, Sorenson, and Seibold 1978; Jablin 1981; Ruback, Dabbs, and Hopper 1984). Although shyness is not associated with brainstorming performance when individuals are tested alone, shy people make fewer contributions than non-shy people in brainstorming groups and hence shy group members generally achieve lower status in brainstorming groups than do non-shy members (Jablin and Sussman 1978; Jablin, Sorenson, and Seibold 1978; Jablin 1981; Comadena 1984). Bruce (1974) found that although girls scored significantly higher than boys on the verbal creativity measures of the Torrance Tests of Creative Thinking when they were tested individually, these girls and boys did not differ either in the number of ideas contributed or in their self-ranked or peer-ranked status when they worked in mixed-sex brainstorming groups. Finally, and contrary to hypotheses concerning paralinguistic status cues, one study found that status within brainstorming groups was not correlated with vocal interruption patterns (Ruback et al. 1984).

Non-Interacting Judges' Evaluations of Original Products

Specific Status Characteristics

Nineteen experiments were found in which subjects were asked to evaluate creations attributed to a creator high or low in domain-specific status (see Table 1). In 16 of the 19 experiments, subjects gave original products significantly more positive evaluations when the products were attributed to creators with high specific status characteristics than when they were attributed to creators with lower or undisclosed specific status characteristics. In two of the 19 experiments, subjects significantly favored the works of higher-status creators on some but not on other dependent measures. In one of the 19 experiments,

Table 1. Experiments on Specific Status Characteristics and the Evaluation of Original Products

Study	Subjects	Type of Product	Dependent Variables	Results
Farnsworth and Beaumont (1929)	64 undergraduates	paintings	bipolar scale with very beautiful and very unattractive poles	significant bias favoring high status creators
Farnsworth and Misumi (1931)	unreported # of undergraduates	paintings	bipolar scale with very beautiful and very unattractive poles	significant bias favoring high status creators on 3 of 8 paintings
Sherif (1935) Study 1	52 undergraduates	prose passages	rank-ordered preference	significant bias favoring high status creators, except for subjects who ignored creators' names
Sherif (1935) Study 2	106 Turkish undergraduates	prose passages	rank-ordered preference	significant bias favoring high status creators
Sherif (1935) Study 3	66 undergraduates	prose passages	rank-ordered preference	significant bias favoring high status creators, except for subjects who ignored creators' names
Michael, Rosenthal, and DeCamp (1949)	120 undergraduates	poems and prose passages	rank-ordered preference	no significant bias
Bernberg (1953)	109 undergraduates	paintings	like vs. dislike	significant bias favoring high status creators
Das, Rath, and Das (1955)	30 Indian undergraduates	poems	rank-ordered preference	significant bias favoring high status creators
Etaugh and Sanders (1974)	210 undergraduates	paintings	emotional impact and overall quality and content	significant bias favoring high status creators on emotional impact, but no significant differences on overall quality and content
Dash (1975)	20 Indian teachers	student essays	grades	significant bias favoring high status creators
Babad (1977)	40 Israeli principals and 70 Israeli undergraduates	essays	composite of interest, profundity, originality, enjoyment, and value to Professionals	significant bias favoring high status creators
Peck (1978)	136 undergraduates	essays	informativeness, enjoyability, clarity and organization, and value to professionals	significant bias favoring high status creators on each measure
Chase (1979)	62 graduate students	essays	grades	significant bias favoring high status creators

Valasek, Avolio, and Forbringer (1979)	135 nursing students	essay	composite of grade and ratings of *persuasiveness,* value to professional readers, value to general readers, stylistic effectiveness, and *profundity,* and of author's *professional competence, professional status,* and *ability to sway*	significant bias favoring high status female creators; no significant bias for male creators
Gibb (1983)	6 undergraduates	essays	grades	significant bias favoring high status creators
Hughes, Keeling, and Tuck (1983)	224 undergraduates	essay	grades	significant bias favoring high status creators
Gillies and Campbell (1985)	34 Scottish undergraduates	poems	bipolar good-bad ratings	significant bias favoring high status creators
Graham and Dwyer (1987)	44 undergraduates	essays	grades	significant bias favoring high status creators, but no significant bias when subjects were intensively trained
Sachdev and Bourhis (1987)	120 Canadian undergraduates	titles for paintings	Tajfel matrices	significant bias favoring high status creators

no significant differences were found. In none of the 19 experiments did original creations receive significantly more positive evaluations when attributed to lower-status creators than when attributed to higher-status creators.

Diffuse Status Characteristics

*Creator's Age.*In three published experiments, researchers manipulated the apparent age of creators. No significant effect on evaluation was found for creator's age in any experiment (see Table 2).

*Creator's Physical Attractiveness.*In 12 published experiments, researchers manipulated the supposed physical attractiveness of creators and then solicited evaluations of their creations (see Table 3). Of these 12 experiments, five provide full support and four provide partial support for the hypothesis that the creator's physical attractiveness favorably biases judges' evaluations of the creator's work. Three of the experiments show no significant effect of physical attractiveness; no study shows a significant negative effect of physical attractiveness.

In these experiments, evaluations made by both male and female judges were biased by the physical attractiveness of both male and female creators. The evidence is stronger, however, for the effects of female creators' physical attractiveness than for that of male creators: seven of the 11 experiments that presented female creators found evidence of a biasing effect of female attractiveness, whereas two of the four studies that presented male creators found evidence of a biasing effect of male attractiveness. Further, there is stronger evidence for male subjects' susceptibility to physical attractiveness than for female subjects' susceptibility: nine of the 11 experiments that used male subjects found evidence of male bias, whereas only three of the eight experiments that used female subjects found evidence of female bias. These patterns are consistent with general findings that the influence of female physical attractiveness is greater than that of male physical attractiveness, and that males are more powerfully influenced by physical attractiveness than are females (Feingold 1990).

*Creator's Race and Ethnicity.*In seven experiments, subjects evaluated original products whose creators were presented as members of different racial or ethnic groups (see Table 4). The experiments do not provide clear evidence of the influence of the creator's race or ethnicity. Of the seven experiments, three showed significant favoritism toward creators belonging to higher-status racial or ethnic groups, three experiments showed no significant effect of the creator's race, and one experiment showed significant favoritism toward creators belonging to a lower-status race.

Table 2. Experiments on Age Discrimination in the Evaluation of Original Products

Study	Subjects	Manipulation	Type of Product	Dependent Variables	Results
Drevenstedt (1981)	136 undergraduates	25 vs. 64 yrs old	essay	overall quality and informational quality	no significant bias
Walsh and Connor (1979)	74 undergraduates	25 vs. 64 yrs old	lithograph	overall quality	no significant bias
Puckett, Petty, Cacioppo, and Fischer (1983)	220 undergraduates	21 vs. 68 yrs old	essays	composite of creativity, ideas, style, writing, arguments, and persuasiveness	no significant bias

189

Table 3. Experiments on Physical Attractiveness and the Evaluation of Original Products

Study	Subjects	Type of Product	Dependent Variables	Results
Landy and Sigall (1974)	60 male undergraduates	essays	general quality and composite of style, creativity, and ideas	significant bias favoring high status creators
Murphy and Hellkamp (1976)	32 undergraduates	paintings	ratings on scale with extremely poor and excellent poles	significant bias favoring high status creators
Fugita, Panek, Balascoe, and Newman (1977)	142 undergraduates	essays	composite of grade and value to general readers, value to professionals, and profundity, authors' professional status, professional competence, view and stylistic effectiveness, and subject's agreement with author	significant bias favoring high status creators
Anderson and Nida (1978)	288 undergraduates	essays	general quality and composite of style, creativity, and ideas	significant bias favoring high status creators
Kaplan (1978) Study 1	140 undergraduates	essays	creativity, style, ideas, and general quality	on each measure, significant bias of male subjects favoring high status creators; no significant bias of female creators
Kaplan (1978) Study 2	120 undergraduates	essays	creativity, style, ideas, and general quality	no significant bias
Bull and Stevens (1979)	48 teachers and 24 adult students	essays	creativity, style, ideas, and general quality	no significant bias
Maruyama and Miller (1980: Study 1)	183 undergraduates	essays	composite of essay creativity, ideas, style, and general quality, and author intelligence, sensitivity, talent, overall ability, and college aptitude	significant bias favoring high status creators

Study	Sample	Material	Measures	Results
Maruyama and Miller (1980: Study 2)	278 undergraduates	essays	composite of essay creativity, ideas, style, and general quality, and author intelligence, sensitivity, talent, overall ability, and college aptitude	significant bias favoring high status creators
Maruyama and Miller (1980: Study 3)	69 undergraduates	essays	composite of essay creativity, ideas, style, and general quality, and author intelligence, sensitivity, talent, overall ability, and college aptitude	no significant bias
Holahan and Stephan (1981)	255 undergraduates	essays	composite of creativity, ideas, general quality, and style	significant bias of male subjects favoring high status creators; no significant bias for female subjects
Cash and Trimer (1984)	216 undergraduates	essays	grades, ratings of general quality, and composite of creativity, style, and ideas	significant bias favoring high status creators on grades and general quality; no significant bias on composite measure

Table 4. Experiments on Racial and Ethnic Discrimination in the Evaluation of Original Products

Study	Subjects	Manipulation	Type of Product	Dependent Variables	Results
Noel and Allen (1976)	184 white nonstudents	Anglo vs. Hispanic	essays	quality	significant bias favoring high status creators
Maruyama and Miller (1980: Study 1)	183 white undergraduates	white vs. black	essays	composite of essay creativity, ideas, style, and general quality, and author intelligence, sensitivity, talent, overall ability, and college aptitude	no significant bias
Maruyama and Miller (1980: Study 2)	278 white undergraduates	white vs. black	essays	composite of essay creativity, ideas, style, and general quality, and author intelligence, sensitivity, talent, overall ability, and college aptitude	no significant bias
Maruyama and Miller (1980: Study 3)	69 white undergraduates	white vs. black	essays	composite of essay creativity, ideas, style, and general quality, and author intelligence, sensitivity, talent, overall ability, and college aptitude	no significant bias
Guttman and Bartal (1982: Study 2)	59 Israeli Sephardic and Ashkenazi undergraduates	Sephardic vs. Ashkenazi	essays	grades	significant bias favoring high status creators, by subjects of both groups
Kruglanski and Freund (1983: Study 2)	39 Sephardic and 105 Ashkenazi Israeli undergraduates	Sephardic vs. Ashkenazi vs. unidentified	essay	literary excellence	significant bias favoring high status creators, by subjects of both groups
Fajardo (1985)	160 white teachers	white vs. black	essays	composite of style, focus, interest, content, spelling, vocabulary, and punctuation	significant bias favoring low status creators

192

Creator's Gender. By far the largest experimental literature on discrimination pertains to gender. Does a creator's gender affect how others evaluate the original product? In part the answer should depend on the domain to which the original product belongs because, as noted earlier, gender stereotypes do not uniformly favor males: whereas males are widely stereotyped as more scientific and mathematical, females are stereotyped as more artistic and literary (Fernberger 1948; Ward and Balswick 1978; see Eagly and Mladinic in press; Simonton 1992b). Further, as noted earlier, it is unclear whether these stereotypes refer to *creative* accomplishment; perhaps, for example, women are perceived as consuming literature more frequently than men, or appreciating literature more sensitively, rather than as writing creative literature more frequently or writing literature more creatively than men. Moreover, gender-related expectations of creative behavior may depend on the degree of creativity; even if females are stereotyped as more creative in art and literature, it seems questionable whether females are generally expected to produce more *world-class* literary or artistic creations. Additionally, expectations for the creative behavior of girls versus boys may differ from those of women versus men.

The most extensive meta-analytic review yet of experimental sexism research was conducted by Swim, Borgida, Maruyama, and Myers (1989). They found only a weak overall bias against females, thus challenging earlier reviews of sexism research (Tosi and Einbender 1985; Wallston and O'Leary 1981). Swim and colleagues' review is difficult to interpret, however, because their selection criteria excluded studies in which subjects rated stimulus persons with little or no individuating information. Research on sexism, including that of Swim and colleagues, has shown consistently that individuating information significantly reduces or eliminates the impact of a stimulus person's gender (e.g., Kasof 1993; Krueger and Rothbart 1988; Locksley, Borgida, Brekke, and Hepburn 1980; Locksley, Hepburn, and Ortiz 1982; Tosi and Einbender 1985). Therefore the unintended effect of reviewing only those studies in which stimulus persons were presented with individuating information was to restrict the review to studies in which sexism effects would be least likely and weakest. Consequently Swim et al.'s review probably underestimated the impact of gender stereotypes on evaluations.

Swim et al.'s (1989) meta-analysis is particularly difficult to interpret as a test of the present hypothesis that the creator's gender affects how others evaluate the original product. In the great majority of studies reviewed by Swim et al., the stimuli that subjects evaluated were not original products. Among the various categories of stimuli analyzed were "applications and essays" and "written works," which included articles, memos, speeches, and artwork. The overlap between "essays" in the first category and "articles" in the second is problematic, as is Swim et al.'s classification of artwork and speeches as "written works." For present purposes, so is the lumping together of essays, which are original products, with applications, which are not original products. In addition, because Swim et al. combined studies in which stereotypes favor female creators (e.g., artists and authors of essays on feminine topics) with studies in which stereotypes favor male creators (e.g., business managers and authors of essays on auto mechanics), it remains possible that discrimination against females on some tasks canceled out discrimination against males on others, thus creating the misleading impression that little discrimination occurred.

These reservations stated, Swim et al.'s (1989) results suggest that judges may have discriminated against female creators' work. Swim et al. meta-analyzed six studies in which "applications and essays" were judged and 28 studies in which "written work" was judged, covering 16 and 86 findings respectively. In most of these studies, however, the primary authors reported data insufficiently to allow calculation of effect sizes. When the analysis is restricted to those findings for which effect sizes were calculable, only nine findings remain under "applications and essays" and 20 under "written work." From these studies, the mean effect sizes are -.23 for "applications and essays" and -.14 for "written work," an indication that male stimulus persons indeed were slightly favored over female stimulus persons. Swim et al. did not report mean effect sizes for these categories when they used study as the unit of analysis and excluded studies in which effect sizes could not be calculated. Even so, there is reason to believe that such results would have shown even greater evidence of sexism because main effect sizes were generally larger when the study rather than the finding was the unit of analysis and when the authors excluded studies in which effect sizes were incalculable. (Swim et al. assigned effect sizes of 0 to the latter studies.)

Further complicating the interpretation of this research, however, is the fact that most experiments on sex discrimination contained a serious methodological flaw. Following Goldberg (1968), most researchers operationalized the creators' gender through the use of sex-typed forenames such as *John* versus *Joan*, *Stephen* versus *Stella*, and *Robert* versus *Roberta*. Although these researchers assumed that sex-typed names vary dichotomously, being either female or male, such names may also differ in attractiveness and connote differential impressions of age, intellectual competence, and social class, each of which may affect evaluations of original products. Several experiments have demonstrated, for instance, that evaluations of essays are affected directly by the attractiveness of the alleged author's forename (Erwin and Calev 1984; Harari and McDavid 1973).

A recent study shows that the experimental literature on sexism was confounded pervasively by forename characteristics other than gender. Kasof

(1993) analyzed the age connotations, intellectual-competence connotations, and attractiveness of the sex-typed forenames used in 144 sexism experiments. It was found that researchers generally used male names that were more attractive, more youthful, and more intellectually competent in their connotations than the female names with which they were contrasted. Moreover, this bias in the naming of stimulus persons correlated highly with the degree of apparent "sexism" reported in a sample of the sexism studies. Much of what appeared to be sexism was in fact discrimination on the basis of forename characteristics other than sex.[5]

Despite this confounding, the experiments uncovered in the present literature search show little evidence of bias against creations attributed to female creators. Of the 44 published experiments listed in Table 5, 30 showed no significant main effect of creator's gender on any dependent variable, 14 showed a significant main effect favoring male creators on at least one dependent variable, and two showed a significant main effect favoring female creators on at least one dependent variable. Of the 138 dependent variables used in these 44 experiments, no significant effect of creator's gender was found on 114 variables (82.6%), significant effects favoring male creators were found on 16 variables (11.6%), and significant effects favoring female creators were found on eight variables (5.8%). Thus the findings of this review are consistent with Swim et al.'s (1989) overall meta-analysis and with Top's (1991) narrative review of sex bias in the evaluation of creations.

Creator's Socioeconomic Status. In five experiments, subjects were led to believe that the creator was of high or low SES and were asked to evaluate their creations (see Table 6). Each study revealed significant bias in favor of high-SES creators. In several of these studies, however, SES was confounded with ethnicity, which leaves open the possibility that judges discriminated on the basis of ethnicity rather than SES.

In four other experiments the creator's status was manipulated by placing advanced degrees next to the creator's name. As shown in Table 6, this manipulation did not cause significant discrimination. Several methodological problems, however, cast doubt on this result. In each study, subjects were presented with scholarly articles that supposedly had been submitted or accepted for publication in academic journals. Hence even when authors were named without advanced degrees, subjects may have assumed reasonably that the authors held such degrees because most academic journals publish articles without noting the authors' degrees.

Table 5. Experiments on Sex Discrimination in the Evaluation of Original Products

Study	Subjects	Type of Product	Dependent Variables	Results
Goldberg (1968)	40 female undergraduates	essays	composite of essays' value to professional readers, value to general readers, stylistic effectiveness, profundity, and persuasiveness, grade, and author's ability to sway, professional competence, and professional status	significant bias favoring high status creators on essays about law, city planning, and linguistics; no significant bias on other essays
Pheterson et al. (1971)	120 female undergraduates	paintings	overall quality and content and emotional impact	no significant main effect of creator status
Baruch (1972)	86 female undergraduates	essays	composite of essays' originality and creativity, value for the general reader, value for the professional reader, logic, persuasiveness, and grade, and authors' professional status and professional competence	no significant main effect of creator status
Chobot et al. (1974)	27 male and 31 female undergraduates	essays	grades, ratings of essays' value to professional readers, value to general readers, stylistic effectiveness, profundity, and persuasiveness	no significant main effects of creator status
Etaugh and Sanders (1974)	105 female and 105 male undergraduates	paintings	overall quality and content and emotional impact	no significant main effects of creator status

196

Study	Sample	Stimulus	Dependent measures	Results
Goldberg (1974)	undisclosed # of males	essays	composite of essays' value to professional readers, value to general readers, stylistic effectiveness, profundity, and persuasiveness, grade, and authors' ability to sway, professional competence, and professional status	bias favoring high status creators, but significance levels not reported
Mischel (1974) Study 1	28 female and 28 male undergraduates and high-school students	essays	composite of grade and ratings of essays' persuasiveness, stylistic effectiveness, rationality, profundity, and value to general readers, and authors' ability to sway, professional competence, and professional status	no significant main effect of creator status; subjects of both sexes significantly favored male authors on masculine essay topics and female authors on feminine essay topics
Mischel (1974) Study 2	27 female and 26 male Israeli kibbutz and high-school students	essays	composite of grade and ratings of essays' persuasiveness, stylistic effectiveness, rationality, profundity, and value to general readers, and authors' ability to sway, professional competence, and professional status	no significant main effects of creator status
Starer and Denmark (1974)	88 female and 88 male undergraduates	poems	rank-ordered preference	no significant main effect; when in group, no significant biases, but when alone males significantly favored low status creators
Etaugh and Rose (1975)	42 female and 42 male students of various levels	essays	grades, ratings of overall quality and content, value, and interest	no significant main effects of creator status
Levenson et al. (1975: Study 1)	55 female and 79 male undergraduates	essays	value and style	no significant main effects of creator status

(continued)

Table 5. (Continued)

Study	Subjects	Type of Product	Dependent Variables	Results
Noel and Allen 1976)	91 female and 93 male adult students	essays	quality	significant bias favoring high status creators
Panek et al. (1976)	57 female and 39 male undergraduates	essays	composite of grade and ratings of essay profundity, stylistic effectiveness, value to professional readers, value to general readers, and persuasiveness, and authors' professional competence and professional status	no significant main effect; subjects of both sexes significantly favored male authors on essays about masculine topics and female authors on essays about feminine topics
Cline et al. (1977)	42 females and 42 males	drawings	bipolar like-dislike ratings	no significant main effect of creator status
Tanner (1977)	56 male and 41 female junior high students	children's story about grizzly bear	composite of whether story was very good, whether author knows much about bears, and whether subject wants to hear another story	significant bias favoring high status creators, in subjects of both sexes
Anderson and Nida (1978)	144 undergraduates of both sexes	essays	general quality and composite of interest, creativity, and style	no significant main effect of creator status
Honig and Carterette (1978)	47 female undergraduates	paintings	overall quality and content and emotional impact	no significant main effects of creator status
Peck (1978)	65 male, 71 female undergraduates	essays	clarity and organization, value to professionals, informativeness, and enjoyability	no significant main effects of creator status
Ward (1978)	80 female undergraduates	essay	style, content, profundity, and persuasiveness	significant bias favoring high status creators on persuasiveness; no other main effects of creator status

Study	Subjects	Stimuli	Variables	Results
Friend et al. (1979)	100 male English undergraduates	essays	composite of overall quality, authors' competence and future success, and subject's willingness to accept author's advice	significant bias favoring high status creators
Valasek et al. (1979)	135 female nursing students	essay	composite of grade and ratings of value to professional readers, value to general readers, stylistic effectiveness, profundity, and persuasiveness, and of author's professional competence, professional status, and ability to sway	significant bias favoring high status creators
Walsh and Connor (1979)	37 undergraduates of each sex	essays	overall quality	no significant main effect of creator sex
Ward (1979)	80 female and 108 male English undergraduates	paintings	composition, use of color, technique, subject matter, warmth, sensitivity, originality, expressiveness, intensity, vitality, overall quality, artistic appeal, artistic competence, and artistic potential	significant bias favoring low status creators on 1 painting's composition, use of color, originality, artistic appeal, and artistic potential; no significant main effects of creator status on other painting
Gross and Geffner (1980)	64 undergraduates of each sex, 48 adult nonstudents of each sex	essays	composite of essay quality, informativeness, and interest, and author's knowledge, qualifications, and career enjoyment	no significant main effect of creator status
Toder (1980)	55 male and 116 female undergraduates	essays	composite ratings of 10 unspecified variables	significant bias favoring high status creators in subjects of both sexes, but only when evaluations were made in mixed-sex groups
Drevenstedt (1981)	73 female and 53 male undergraduates	essays	overall quality and informational quality	no significant main effect of creator sex

199

(continued)

Table 5. (Continued)

Study	Subjects	Type of Product	Dependent Variables	Results
Ellerman, Dowling, Hihschen, Kemp, and White (1981)	194 female and 77 male Australian teachers and trainees	poems and paintings	creativity, overall quality and content, and emotional impact	no significant main effect of creator status
Etaugh and Kasley (1981)	184 undergraduates of each sex	essay	grades and ratings of value, style, and persuasiveness	significant bias favoring high status creators on each dependent variable
Isaacs (1981)	178 undergraduates, number of each sex unstated	essays	composite of grade and ratings of essays' value to professional readers, value to general readers, stylistic effectiveness, profundity, and persuasiveness, and author professional competence and professional status	no significant main effect of creator status
Ward (1981) Study 1	58 male undergraduates	essay	style, content, professionalism, persuasiveness, and profundity	no significant main effect of creator status
Ward (1981) Study 2	33 female undergraduates	essay	style, content, professionalism, persuasiveness, and profundity	no significant main effect of creator status
Ward (1981) Study 3	32 female and 60 male English undergrads; 44 female and 50 male art college students	painting	composition, use of color, technique, subject matter, warmth, sensitivity, originality, expressiveness, intensity, vitality, overall quality, artistic appeal, artistic competence, and artistic potential	significant bias favoring high status creators on composition, technique, expressiveness, overall quality, and artistic appeal in art students; no significant main effects of creator sex in undergraduates

Study	Sample	Stimulus materials	Dependent measures	Results
Ellerman and Smith (1983)	11 male and 124 female Australian undergraduates	paintings	composite of creativity, overall quality and content, emotional impact, and creator's technical competence and artistic future	no significant main effect of creator status
Etaugh and Foresman (1983)	80 undergraduates of each sex	essay	grade and ratings of value, persuasiveness, and writing style	no significant main effect of creator status
Lenney et al. (1983: Study 2)	57 male and 61 female undergrads	paintings and pottery	composite of overall quality, creativity, and attractiveness, and creator's technical competence, creativity, chances of selling work, and chances of winning prize	no significant main effect of creator status
Paludi and Bauer (1983)	180 undergraduates of each sex	essays	composite of value, style, persuasiveness, ability to sway reader, and intellectual depth, and creator's professional status and knowledge of the field	significant bias favoring high status creators, in subjects of both sexes
Cash and Trimer (1984)	216 female undergraduates	essays	grades, rating of general quality, and composite of creativity, ideas, and style	no significant main effect of creator status
Lipton and Hershaft (1984)	60 undergraduates of each sex	paintings	willingness to display painting in one's home and emotionally moving	significant bias favoring high status creators on emotionally moving in subjects of both sexes; bias on willingness depended on how creator was identified
Haemmerlie et al. (1985)	80 male undergraduates	essays	value, profundity, persuasiveness, and writing style	no significant main effect of creator status

(continued)

Table 5. (Continued)

Study	Subjects	Type of Product	Dependent Variables	Results
Paludi and Strayer (1985)	150 undergraduates of each sex	essays	composite of value, style, quality persuasiveness, and depth and insight, and author's professional competence, ability to sway, professional status, and knowledge of the field	significant bias favoring high status creators
Aamiry and Stertich (1986)	34 Jordanian undergraduates of each sex	essays	composite of value, style, profundity, and persuasiveness, and authors' professional competence, ability to sway, and professional status	no significant main effect of creator status
Etaugh, Houlter, and Ptasnik (1988)	78 male, 78 female undergraduates	essay	grade and ratings of value and writing style	no significant main effect of creator status
Guttman and Boudo (1988)	23 male, 218 female Israeli teachers	essay	grade	significant bias favoring high status creator
Rickard (1990)	145 female undergraduates	paintings	creativity, suitability, technical competence, and overall quality	significant bias favoring high status creators on technical competence but favoring low status creators on creativity, overall quality, and suitability

202

Table 6. Experiments on Socioeconomic Status and the Evaluation of Original Products

Study	Subjects	Manipulation	Type of Product	Dependent Variables	Results
Babad (1977)	40 Israeli principals 72 Israeli teaching students	Harvard professor vs. Sephardic	essays	composite of depth, interest, enjoyment, professional worth and originality	significant bias favoring high status creators
Peck (1978)	136 undergraduates	associate professor with many awards and publications vs. doctoral candidate	essays	value to professionals, informativeness, clarity, and enjoyability	significant bias favoring high status creators
Guttman and Bartal (1982, Study 2)	59 Israeli undergraduates, about half Sephardic and half Ashkenazi	Sephardic vs. Ashkenazi	essays	grades	significant bias favoring high status creators, in both groups of subjects
Kruglanski and Freund (1983, Study 2)	39 Sephardic and 105 Ashkenazi Israeli teaching students	Sephardic vs. Ashkenazi vs. unidentified	essay	literary excellence	significant bias favoring high status creators, in both groups of subjects
Van Oudenhaven, Siero, and Withag (1984: Study 2)	unknown	middle-SES vs. low-SES	essays	grades	significant bias favoring high status creators
Panek et al. (1976)	96 undergraduates	advanced degrees vs. no degrees	essays	composite of grade and ratings of stylistic effectiveness, profundity, value to professionals, value to general readers, and persuasiveness, and authors' professional competence and professional status	no significant effect of creator status

(continued)

Table 6. (Continued)

Study	Subjects	Manipulation	Type of Product	Dependent Variables	Results
Valasek et al. (1979)	135 undergraduates	nursing degree vs. no degree	essay	composite of grade and ratings of stylistic effectiveness, profundity, value to professionals, value to general readers, and persuasiveness, and author's professional competence, professional status, and ability to sway	no significant main effect of creator status
Isaacs (1981)	178 undergraduates	title of address (Dr. vs. Mr. and Miss)	essays	composite of grade and ratings of value to professionals, value to general readers, stylistic effectiveness, profundity, and persuasiveness, and authors' professional competence and professional status	no significant main effect of Dr. title
Haemmerlie et al. (1985)	80 undergraduates	advanced degrees vs. no degrees	essays	value, persuasiveness, profundity, and style	no significant main effects of creator status

Third-Party Status Characteristics

Four experiments are relevant to the hypothesis that a third party's specific tatus characteristics affect evaluations of original products. As revealed in Table 7, each of these experiments found evidence for the predicted effects of third-party status characteristics on the evaluation of original creations.

Some Limitations of the Experimental Literature

This review provides mixed support for the hypothesis that the creator's status characteristics influence how others evaluate the creator's original products. Experiments on group brainstorming, specific status characteristics, third-party status effects, and the diffuse status characteristics of physical attractiveness and socioeconomic status provided consistent support for the status characteristics hypotheses, whereas experiments on the diffuse status characteristics of age and gender produced few significant effects.

This experimental literature has several important limitations. Each study used essentially the same experimental paradigm, that established by Goldberg (1968), in which subjects are presented with a creation to evaluate. Some subjects are informed that the creator has a positive status characteristic; others are given to believe that the creator has a lower value of the same status characteristic. In Goldberg's classic experiment, for example, half of the subjects evaluated an essay whose author was identified by a female name and title of address, while the other half evaluated an identical essay whose author was identified by a male name and title. Because the great majority of studies cited in this review used between-subjects designs rather than within-subjects designs, the salience of the alleged creators' status characteristics may have been low, and therefore the experimental judgment-tasks may satisfy neither the scope conditions of status characteristics theory nor the scope-extension proposed earlier in this article. Further, because the researchers generally did not perform manipulation checks, the effectiveness of such subtle experimental manipulations is open to doubt.

Another shortcoming in most of the experiments reviewed here concerns composite measures of discrimination. As noted previously, social groups are not stereotyped as uniformly positive or negative but rather as competent at certain tasks and incompetent at others. For example, Blacks are stereotyped as intellectually inferior but athletically superior to whites, and females are stereotyped as less technically competent but more artistic and more literary than males. These differences lead to differential expectation (dis)advantages depending on the type of task; for example, males have been found to benefit from an expectation advantage over females on masculine-typed tasks whereas

Table 7. Experiments on Third-Party Status Effects on the Evaluation of Original Products

Study	Subjects	Manipulation	Type of Product	Dependent Variables	Results
Moore (1921)	95 subjects	opinion of music expert vs. no opinion	music	ranked preference	significant bias favoring expert opinion
Bernberg (1953)	109 undergraduates	opinions of prominent art critics vs. no opinion	paintings	like vs. dislike	significant bias favoring expert opinion
Cole (1954)	40 undergraduates	opinions of art professor favoring different paintings	paintings	ranked preference	significant bias favoring expert opinion
Aronson, Turner, and Carlsmith (1963)	112 undergraduates	critical essay attributed to T. S. Eliot vs. undergraduate	poems	rank-order of "the way the poet uses form to aid in expressing his meaning"	significant bias favoring high status critic

females benefit from an expectation advantage over males on feminine-typed tasks (Carli 1991; Dovidio, Brown, Heltman, Ellyson, and Keating 1988; Eagley, Makhijani, and Klonsky 1992). Following Goldberg (1968), however, many researchers ignored this complexity and instead combined into composite scores the subjects' ratings of, for example, the technical competence and the creativity of the creation. If females are thought to be less technically competent but more artistically talented than males, then ratings of technical competence favoring males might cancel out ratings of creativity or overall quality favoring females; the resulting composite measure would give the erroneous impression that little or no discrimination occurred. Support for this notion comes from the few studies in which these judgments were analyzed separately. Rickard (1990), for example, found that when paintings were attributed to male artists, they were judged to be significantly higher in technical competence but significantly lower in creativity and in overall quality than when they were attributed to female artists (also see Pheterson, Kiesler, and Goldberg 1971). Additionally, many sexism studies combined into a single dependent measure judgments of performance on masculine-typed creativity tasks and on feminine-typed creativity tasks.

Composite measures typically included guesses about the creator's professional status or predictions of the creator's future career success. Such judgments, however, may simply reveal the subjects' beliefs about how much discrimination the creator faces from others, rather than reveal the subjects' own prejudices. A prediction of greater career success for males and whites, for example, may reflect a belief that females and Blacks face greater discrimination in their careers than do males and whites, rather than a belief that females and Blacks are less competent than males and whites. This notion is supported by Rickard's (1990) study, which found that although subjects judged paintings attributed to female artists to be higher in creativity and in overall quality than those attributed to male artists, the subjects predicted greater career success for male artists than for female artists.

Potential confounding by in-group favoritism is another general shortcoming of the experimental literature. Individuals who have ordinary self-serving biases tend to identify more with successful others than with unsuccessful others (Cialdini, Borden, Thorne, Walker, Freeman, and Sloan 1976; Pleban and Tesser 1981; Snyder, Lassegard, and Ford 1986). Because the experiments reviewed earlier presented subjects with creators high or low in a widely valued characteristic, subjects may have identified more with higher-status creators than with lower-status creators. Once so identified, self-serving subjects should be biased in favor of original products made by the higher-status creators (Kasof in press). If this is the case, then what appears to be purely cognitive discrimination based on performance expectations may actually be a motivational bias resulting from self-serving tendencies in social identification and evaluation.

A final general shortcoming to be noted concerns the limited range of stimulus creations. Because most researchers used essays as the creations to be judged, it is unclear whether status has the same effects on evaluations of other kinds of creations, such as poems, recipes, and scientific theories. Moreover, the essays that researchers used spanned a very narrow range of originality, difficulty, length, and so forth. Researchers generally used essays low in originality, fewer than 700 words long, and concerned with such mundane topics as "How to Buy a Used Motorcycle" and "Our Lady of the Bicentennial." The narrowness of this range of stimulus characteristics further limits the generality of conclusions that can be drawn from these studies.

CONCLUSION

Historically, creativity research has been conducted primarily by personality psychologists and cognitive psychologists examining the dispositional characteristics of successful creators. Long neglected have been the social contexts in which creators and potential creators are embedded; most neglected of all have been the larger-scale sociological forces that impinge profoundly, though often imperceptibly, on the creative process. Why does creativity vary by race, gender, social class, and nationality? Why are the pages of prestigious encyclopedias filled so disproportionately with the names of white male creators from high-SES families? Is it simply that innate talent is distributed unevenly across social groups? The traditional approaches to creativity have not provided satisfying answers to these questions; indeed, they have seldom even addressed such issues.

In this article, I have departed from the traditional approach by focusing on how status processes affect the evaluation of original products. Creative products, by definition, are products that are both original and positively evaluated. Thus creativity, far from being a purely objective attribute inherent to the creative product, necessarily involves subjective judgment and interaction between creator, product, and judge. Status processes that influence the evaluation of original products thereby influence the level of creativity the products attain and thus constitute an external influence on creativity.

To situate the present effort in the context of past and future creativity research, it may be useful to think of creativity as a form of persuasive communication, in which the creator is the *source*, the original product is the *message*, and the judge is the *audience* (see Figure 1).[6] Just as a message cannot be persuasive without an audience that is persuaded, neither can an original

creation		evaluation
Creator ------------------>	Original Product <-------------------	Judge
(Source)	(Message)	(Audience)
Dispositional Influences		Dispositional Influences
External-Nonsocial Influences		Exernal-Nonsocial Influences
Micro-social Influences		Micro-social Influences
Macro-social Influences		Macro-social Influences

Figure 1. Multiple Causal Infleunces on the
Creation and the Evaluation of Original Products

product be creative unless it is positively accepted by an audience; and just
as persuasion is caused by multiple factors both internal and external to the
source of the persuasive communication, so creativity is caused by multiple
factors both internal and external to the maker of a creative product. Indeed,
both the creation and the evaluation of original products may be influenced
by dispositional, external-nonsocial, micro-social, and macro-social causes.
For example, the creation of original products may be influenced not only by
the creator's dispositions but also by noise in the creator's immediate
surrounding (Kasof 1992, chap. 2), exposure to salient external contingencies
(Amabile 1983), and large-scale historical forces within a domain or genre (Li
and Gardner 1993; Martindale 1975). Similarly, the evaluation of original
products may be influenced by the judge's integrative complexity (Tetlock
1988), in-group favoritism (e.g., Brown, Schmidt, and Collins 1988; Ferguson
and Kelley 1964), and national social mobility (Jorgenson 1975; Kasof 1992,
chap. 5; Sales 1973). Clearly, the dispositionally-oriented, creator-centered
approach that has dominated creativity research since the field's inception can
explain only a small fraction of the many important influences on creativity.
Just as clearly, a great deal of research, conducted by scientists in many different
fields, remains yet to be done.

ACKNOWLEDGMENT

I gratefully acknowledge Joseph Berger, Roger Brown, Stanley Lieberson, Barry
Markovsky, David Riesman, Steven Rytina, and Dean Simonton, and an anonymous
reviewer for helpful comments on an earlier draft of this paper. This paper is a revised
version of Chapter Four of my doctoral dissertation, *Social Determinants of Creativity,*
Harvard University, November 1992.

NOTES

1. For some sociological research relevant to creativity, see Becker (1982), Bourdieu (1984), Brannigan (1981), Cole (1992), Gans (1974), Greenfeld (1989), Sorokin (1937-1941), and White (1993).

2. Although members of brainstorming groups are instructed to avoid judging ideas (Osborn 1957), comparative evaluation of others' versus one's own ideas nevertheless is common in brainstorming groups (Collaros and Anderson 1969; Diehl and Stroebe 1987; Jabin 1981; Paulus and Dzindolet 1993; Paulus, Dzindolet, Poletes, and Camacho 1993).

3. Situations satisfying these extended scope conditions are not limited to creativity judgments. For example, an eyewitness's task of identifying a criminal from a police line-up ordinarily is neither interactive nor collective. According to the new scope-extension, if the persons in the line-up vary in status characteristics such as race, age, or physical attractiveness, the eyewitness's judgments of the line-up members should be influenced by such status characteristics.

4. One consequence is that creators should prefer that their creations be performed by higher-status rather than lower-status performers. When pop star Madonna asked the struggling songwriter Brian Elliot for permission to record his song "Pappa Don't Preach," he had already promised the song to an unestablihed singer whose first album he had been producing for six months, but he was instead "persuaded by a great many people that to have Madonna cut the song would...bring a lot of credibility...and move my career ahead by four or five years in a six-month span" (Bronson 1992, p. 644).

5. The sexism experiments reviewed by Kasof (1993) include 28 in which original products were used as the stimuli to be judged and in which mismatched forenames were used to identify the creditor's gender; thus any conclusions about sexism based on this literature must be questioned. These 28 studies are Baruch (1972); Cash and Trimer (1984); Chobot, Goldberg, Abramson, and Abramson (1974); Cline, Holmes, and Werner (1977); Etaugh and Foresman (1983); Etaugh and Kasley (1981); Etaugh and Rose (1975); Etaugh and Sanders (1974); Goldberg (1968, 1974); Gross and Feffner (1980); Haemmerlie, Abdul-Wakeel, and Poomeroy (1985); Honig and Carterrette (1978); Isaacs (1981); Lenney, Mitchell and Browning (1983); Levenson, Burford, Bonno, and Davis (1975); Lipton and Hershaft (1984); Mischel (1974); Noel and Allen (1976); Paludi and Bauer (1983); Paludi and Strayer (1985); Panek, Deitchman, Burkholder, Speroff, and Haude (1976); Peck (1978); Pheterson et al. (1971); Starer and Denmark (1974); Tanner (1977); Valasek, Avolio, and Forbringer (1979); Walsh and Connor (1979).

6. Approaching creativity as a form of persuasive communication is consistent with Simonton's (1988) view of creativity as a kind of cultural leadership and with Csikszentmihalyi's (1988) view of creativity as located not in the creator or in the product but rather in the interaction between the creator and the field's gatekeepers who selectiviey retain or reject creators' products.

REFERENCES

Aamiry, A. and D. Steitieh. 1986. "The Impact of Reader's Sex and Training on Gender Prejudice Among University Students in Jordan." *Dirasat* 13: 7-17.

Allison, P. D., and J. S. Long. 1987. "Interuniversity Mobility of Academic Scientists." *American Sociological Review* 52: 643-652.

_____. 1990. "Departmental Effects on Scientific Productivity." *American Sociological Review* 55: 469-478.

Amabile, T. M. 1982. "Social Psychology of Creativity: A Consensual Assessment Technique." *Journal of Personality and Social Psychology* 43: 997-1013.

Amabile, T. M. 1983. *The Social Psychology of Creativity.* New York: Springer-Verlag.

Anderson, R. and S. Nida. 1978. "Effect of Physical Attractiveness on Opposite-and Same-sex Evaluations." *Journal of Personality* 46: 401-413.

Aronson, E., J. Turner, and J. Carlsmith. 1963. "Communicator Credibility and Communication Discrepancy as Determinants of Opinion Change." *Journal of Abnormal and Social Psychology* 67: 31-36.

Babad, E. 1977. "Effect of Source of Information as a Function of Age, Professional Relevance, and Experience." *Psychological Reports* 41: 231-236.

Bakanic, V., C. McPhail, and R. J. Simon. 1987. "The Manuscript Review and Decision-making Process." *American Sociological Review* 52: 631-642.

Baker, P. 1985. "The Status of Age: Preliminary Results." *Journal of Gerontology* 40: 506-508.

————. 1988. "Age and Sex as Diffuse Status Characteristics." *International Journal of Small Group Research*: 76-83.

Barber, B. 1961. "Resistance by Scientists to Scientific Discovery." *Science* 134: 596-602.

Bare, E. A. and G. Ptacek. 1988. *Mothers of Invention: From the Bra to the Bomb, Forgotten Women.* New York: Morrow.

Baruch, G. 1972. "Maternal Influences Upon College Women's Attitudes Toward Women and Work." *Developmental Psychology* 6: 32-37.

Beattie, G., C. Agahi, and C. Spencer. 1982. "Social Stereotypes Held by Different Occupational Groups in Post-revolutionary Iran." *European Journal of Social Psychology* 12: 75-88.

Becker, H. S. 1982. *Art Worlds.* Berkeley and Los Angeles: University of California Press.

Berger, J. and M. Zelditch (Eds.). 1985. *Status, Rewards, and Influence.* San Francisco:Jossey-Bass.

Berger, J., H. Fisek, R. Norman, and M. Zelditch. 1977. *Status Characteristics and Social Interaction: An Expectation States Approach.* New York: Elsevier.

Berger, J., M. Webster, C. Ridgeway, and S. J. Rosenholtz. 1986. "Status Cues, Expectations, and Behavior." Pp. 1-22 in *Advances in Group Processes* (Volume 3), edited by E. Lawler. Greenwich, CT: Jai Press.

Bernberg, R. 1953. "Prestige Suggestion in Art as Communication." *Journal of Social Psychology* 38: 23-30.

Bourdieu, P. 1984. *Distinction: A Social Critique of the Judgment of Taste.* Cambridge, MA: Harvard University Press.

Boyd, J. and J. Dowd. 1988. "The Diffuseness of Age." *Social Behaviour* 3: 85-103.

Brannigan, A. 1981. *The Social Basis of Scientific Discoveries.* Cambridge: Cambridge University Press.

Bronson, F. 1992. *The Billboard Book of Number One Hits.* New York: Billboard Publications.

Brown, J., G. Schmidt, and R. Collins. 1988. "Personal Involvement and the Evaluation of Group Products." *European Journal of Social Psychology* 18: 177-179.

Bruce, P. 1974. "Reactions of Preadolescent Girls to Science Tasks." *Journal of Psychology* 86: 303-308.

Bull, R. and J. Stevens. 1979. "The Effects of Attractiveness of Writer and Penmanship on Essay Grades." *Journal of Occupational Psychology* 52: 53-59.

Carli, L. L. 1991. "Gender, Status, and Influence." Pp. 89-113 in *Advances in Group Processes* (Volume 8), edited by E. Lawler. Greenwich, CT: Jai Press.

Cash, T. and C. Trimer. 1984. "Sexism and Beautyism in Women's Evaluations of Peer Performance." *Sex Roles* 10: 87-98.

Cattell, R. and J. Drevdahl. 1955. "A Comparison of the Personality Profile of Eminent Researchers with that of Eminent Teachers and Administrators." *British Journal of Psychology* 44: 248-261.

Chase, C. 1979. "The Impact of Achievement Expectations and Handwriting Quality on Scoring Essay Tests." *Journal of Educational Measurement* 16: 39-42.

Chobot, D., P. Goldberg, L. Abramson, and P. Abramson. 1974. "Prejudice Against Women: A Replication and Extension." *Psychological Reports* 35: 478.

Cialdini, R., R. Borden, A. Thorne, M. Walker, S. Freeman, and L. Sloan. 1976. "Basking in Reflected Glory: Three (Football) Field Studies." *Journal of Personality and Social Psychology* 34: 366-375.

Clifford, M. and E. Hatfield. 1973. "The Effects of Physical Attractiveness on Teacher Expectations." *Sociology of Education* 46: 248-258.

Cline, M., D. Holmes, and J. Werner. 1977. "Evaluations of the Work of Men and Women as a Function of the Sex of the Judge and Type of Work." *Journal of Applied Social Psychology* 7: 89-93.

Cohen, E. G. and S. S. Roper. 1972. "Modification of Interracial Interaction Disability: An Application of Status Characteristic Theory." *American Sociological Review* 37: 643-657.

Cohen, I. 1985. *Revolution in Science*. Cambridge, MA: Harvard University Press.

Cole, D. 1954. "'Rational Arguments' and 'Prestige Suggestion' as Factors Influencing Judgment." *Sociometry* 17: 350-354.

Cole, J. and S. Cole. 1973. *Social Stratification in Science*. Chicago: University of Chicago Press.

Cole, S. 1992. *Making Science: Between Nature and Society*. Cambridge: Harvard University Press.

Cole, S., L. Rubin, and J. R. Cole. 1979. "Peer Review and the Support of Science." *Scientific American* 237: 34-41.

Collaros, P. A. and L. R. Anderson. 1969. "Effect of Perceived Expertness Upon Creativity of Members of Brainstorming Groups." *Journal of Applied Psychology* 53: 159-163.

Comadena, M. E. 1984. "Brainstorming Groups: Ambiguity Tolerance, Communication Apprehension, Task Attraction, and Individual Productivity." *Small Group Behavior* 15: 251-264.

Crane, D. 1965. "Scientists at Major and Minor Universities: A Study of Productivity and Recognition." *American Sociological Review* 30: 699-714.

————. 1967. "The Gatekeepers of Science: Some Factors Influencing the Selection of Articles for Scientific Journals." *American Sociologist* 32: 195-201.

Crockett, W. and M. Hummert. 1987. "Perceptions of Aging and the Elderly." *Annual Review of Gerontology and Geriatrics* 7: 217-241.

Csikszentmihalyi, M. 1988. "Society, Culture, and Person: A Systems View of Creativity." Pp. 325-339 in *The Nature of Creativity*, edited by R. J. Sternberg. Cambridge: Cambridge University Press.

————. 1990. "The Domain of Creativity." Pp. 190-212 in *Theories of Creativity*, edited by M. A. Runco and R. S. Albert. Newbury Park, CA: Sage.

Csikszentmihalyi, M. and R. Robinson. 1986. "Culture, Time, and the Development of Talent." Pp. 264-284 in *Conceptions of Giftedness*, edited by R. J. Sternberg and J. E. Davidson. New York: Cambridge University Press.

Das, J., R. Rath, and R. Das. 1955. "Understanding Versus Suggestion in the Judgment of Literary Passages." *Journal of Abnormal and Social Psychology* 51: 624-628.

Dash, A. 1975. "A Study of the Expectancy Bias and the Suggestion Bias of Teachers." *Journal of Psychological Researches* 19: 17-19.

Diehl, M. and W. Stroebe. 1987. "Productivity Loss in Brainstorming Groups: Toward the Solution of a Riddle." *Journal of Personality and Social Psychology* 53: 497-509.

DiMaggio, P. 1987. "Classification in Art." *American Sociological Review* 52: 440-455.

DiMaggio, P. and M. Useem. 1978. "Social Class and Arts Consumption: The Origins and Consequences of Class Differences in Exposure to the Arts in America." *Theory and Society* 5: 141-161.

Domino, G. 1970. "Identification of Potentially Creative Persons from the Adjective Check List." *Journal of Consulting and Clinical Psychology* 1: 48-51.

Dovidio, J. F., C. E. Brown, K. Heltman, S. L. Ellyson, and C. F. Keating. 1988. "Power Displays Between Women and Men in Discussions of Gender-linked Tasks: A Multichannel Study." *Journal of Personality and Social Psychology* 55: 580-587.

Dovidio, J., S. Ellyson, C. Keating, K. Heltman, and C. Brown. 1988. "The Relationship of Social Power to Visual Displays of Dominance Between Men and Women." *Journal of Personality and Social Psychology* 54: 233-242.

Drevenstedt, J. 1981. "Age Bias in the Evaluation of Achievement: What Determines?" *Journal of Gerontology* 36: 453-454.

Eagly, A. H. and A. Mladinic. In press. "Are People Prejudiced Against Women? Some Answers from Research on Attitudes, Gender Stereotypes, and Judgments of Competence." In *European Review of Social Psychology*, edited by W. Stroebe and M. Hewstone. New York: Wiley.

Eagly, A. H., R. D. Ashmore, M. G. Makhijani, and L. C. Longo. 1991. "What is Beautiful is Good, but...: A Meta-analytic Review of Research on the Physical Attractiveness Stereotype." *Psychological Bulletin* 110: 109-128.

Eagly, A. H., M. G. Makhijani, and B. G. Klonsky. 1992. "Gender and the Evaluation of Leaders: A Meta-analysis." *Psychological Bulletin* 111: 3-33.

Ellerman, D. and E. Smith. 1983. "Generalized and Individual Bias in the Evaluation of the Work of Women: Sexism in Australia." *Australian Journal of Psychology* 35: 71-79.

Ellerman, D., C. Dowling, M. Hinschen, J. Kemp, and L. White. 1981. "Teachers' Evaluations of Creative Work by Children: Sexism in Australia." *Psychological Reports* 48: 439-446.

Erwin, P. and A. Calev. 1984. "The Influence of Christian Name Stereotypes on the Marking of Children's Essays." *British Journal of Educational Psychology* 54: 223-227.

Etaugh, C. and E. Foresman. 1983. "Evaluations of Competence as a Function of Sex and Marital Status." *Sex Roles* 9: 759-765.

Etaugh, C. and H. C. Kasley. 1981. "Evaluating Competence: Effects of Sex, Marital Status, and Parental Status." *Psychology of Women Quarterly* 6: 196-203.

Etaugh, C. and S. Rose. 1975. "Adolescents' Sex Bias in the Evaluation of Performance." *Developmental Psychology* 11: 663-664.

Etaugh, C. and S. Sanders. 1974. "Evaluation of Performance as a Function of Status and Sex Variables." *Journal of Social Psychology* 94: 237-241.

Etaugh, C., B. Houtler, and P. Ptasnik. 1988. "Evaluating Competence of Women and Men." *Psychology of Women Quarterly* 12: 191-200.

Fajardo, D. 1985. "Author Race, Essay Quality, and Reverse Discrimination." *Journal of Applied Social Psychology* 15: 255-268.

Farnsworth, P., and H. Beaumont. 1929. "Suggestion in Pictures." *Journal of General Psychology* 2: 362-366.

Farnsworth, P. and I. Misumi. 1931. "Further Data on Suggestion in Pictures." *American Journal of Psychology* 43: 632.

Feingold, A. 1990. "Gender Differences in Effects of Physical Attractiveness on Romantic Attraction: A Comparison Across Five Research Paradigms." *Journal of Personality and Social Psychology* 59: 981-993.

————. 1992. "Good-looking People are not What They Seem." *Psychological Bulletin* 111: 304-341.

Feldman, J. 1972. "Stimulus Characteristics and Subject Prejudice as Determinants of Stereotype Attribution." *Journal of Personality and Social Psychology* 21: 333-340.

Ferguson, C. and H. Kelley. 1964. "Significant Factors in Overevaluation of Own-group's Product." *Journal of Abnormal and Social Psychology* 69: 223-228.

Fernberger, S. 1948. "Persistence of Stereotypes Concerning Sex Differences." *Journal of Abnormal and Social Psychology* 43: 97-101.

Fine, G. 1983. "Cheating History: The Rhetorics of Art Forgery." *Empirical Studies of the Arts* 1: 75-93.

Friend, P., R. Kalin, and H. Giles. 1979. "Sex Bias in the Evaluation of Journal Articles: Sexism in England." *British Journal of Social and Clinical Psychology* 18: 77-78.

Fugita, S., P. Panek, L. Balascoe, and I. Newman. 1977. "Attractiveness, Level of Accomplishment, Sex of Rater, and the Evaluation of Feminine Competence." *Representative Research in Social Psychology* 8: 1-11.

Gans, H. J. 1974. *Popular Culture and High Culture*. New York: Basic Books.

Gardner, H. 1993. *Creating Minds*. New York: Basic Books.

Getzels, J. and M. Csikszentmihalyi. 1976. *The Creative Vision: A Longitudinal Study of Problem-finding in Art*. New York: Wiley-Interscience.

Gibb, G. 1983. "Influence of "Halo" and "Demon" Effects in Subjective Grading." *Perceptual and Motor Skills* 56: 67-70.

Gillies, J. and S. Campbell. 1985. "Conservatism and Poetry Preferences." *British Journal of Social Psychology* 24: 223-227.

Goldberg, P. 1968. "Are Women Prejudiced Against Women?" *Trans-Action: Social Science and Modern Society* 5: 28-30.

————. 1974. "Prejudice Toward Women: Some Personality Correlates." *International Journal of Group Tensions* 4: 53-63.

Gough, H. 1961. "Techniques for Identifying the Creative Research Scientist." In *The Creative Person*, edited by D. MacKinnon. Berkeley: University of California Press.

————. 1979. "A Creative Personality Scale for the Adjective Check List." *Journal of Personality and Social Psychology* 37: 1398-1405.

Graham, S., and A. Dwyer. 1987. "Effects of the Learning Disability Label, Quality of Writing Performance, and Examiner's Level of Expertise on the Evaluation of Written Products." *Journal of Learning Disabilities* 20: 317-318.

Greenfeld, L. 1989. *Different Worlds: A Sociological Study of Taste, Choice and Success in Art*. Cambridge: Cambridge University Press.

Gross, M. and R. Geffner. 1980. "Are the Times Changing? An Analysis of Sex-Role Prejudice." *Sex Roles* 6: 713-721.

Guttmann, J. and D. Bar-Tal. 1982. "Stereotypic Perceptions of Teachers." *American Educational Research Journal* 19: 519-528.

Guttmann, J. and M. Boudo. 1988. "Teachers' Evaluations of Pupils' Performance as a Function of Pupils' Sex, Family Type and Past School Performance." *Educational Review* 40: 105-113.

Haemmerlie, F., A. Abdul-Wakeel, and M. Pomeroy. 1985. "Male Sex Bias Against Men and Women in Various Professions." *Journal of Social Psychology* 126: 797-798.

Harari, H. and J. McDavid. 1973. "Name Stereotypes and Teachers' Expectations." *Journal of Educational Psychology* 65: 222-225.

Harrington, D. 1975. "Effects of Explicit Instructions to 'be Creative' on the Psychological Meaning of Divergent Thinking Test Scores." *Journal of Personality* 43: 434-454.

Helmreich, R. L., J. T. Spence, W. E. Beane, G. W. Lucker, and K. A. Matthews. 1980. "Making it in Academic Psychology: Demographic and Personality Correlates of Attainment." *Journal of Personality and Social Psychology* 39: 896-908.

Helson, R. 1971. "Women Mathematicians and the Creative Personality." *Journal of Consulting and Clinical Psychology* 36: 210-211; 217-220.

_____. 1980. *Women and the Mathematical Mystique.* Baltimore: The Johns Hopkins Press.

_____. 1985. "Which of Those Young Women with Creative Potential Became Productive?" In *Perspectives in Personality Theory, Measurement, and Interpersonal Dynamics,* edited by R. Hogan and W. Jones. Greenwich, CT: Jai Press.

Hoffmann, B. 1972. *Albert Einstein: Creator and Rebel.* New York: New American Library.

Holahan, C. and C. Stephan. 1981. "When Beauty Isn't Talent: The Influence of Physical Attractiveness, Attitudes Toward Women, and Competence on Impression Formation." *Sex Roles* 7: 867-876.

Honig, A. and E. Carterette. 1978. "Evaluation by Women of Painters as a Function of their Sex and Achievement and Sex of the Judges." *Bulletin of the Psychonomic Society* 11: 356-358.

Howard, J. and K. Pike. 1986. "Ideological Investment in Cognitive Processing: The Influence of Social Statuses on Attribution." *Social Psychology Quarterly* 49: 154-167.

Hughes, D., B. Keeling, and B. Tuck, B. 1983. "Effects of Achievement Expectations and Handwriting Quality on Scoring Essays." *Journal of Educational Measurement* 20: 65-70.

Isaacs, M. 1981. "Sex Role Stereotyping and the Evaluation of the Performance of Women: Changing Trends." *Psychology of Women Quarterly* 6: 187-195.

Jablin, F. M. 1981. "Cultivating Imagination: Factors that Enhance and Inhibit Creativity in Brainstorming Groups." *Human Communication Research* 7: 245-258.

Jablin, F. M. and L. Sussman. 1978. "An Exploration of Communication and Productivity in Real Brainstorming Groups." *Human Communication Research* 4: 329-337.

Jablin, F. M., R. L. Sorenson, and D. R. Seibold. 1978. "Interpersonal Perception and Groups Brainstorming Performance." *Communication Quarterly* 26: 36-44.

Jellison, J. and J. Riskind. 1970. "A Social Comparison of Abilities Interpretation of Risk-taking Behavior." *Journal of Personality and Social Psychology* 15: 375-390.

Jellison, J., J. Riskind, and L. Broll. 1972. "Attribution of Ability to Others on Skill and Chance Tasks as a Function of Level of Risk." *Journal of Personality and Social Psychology* 22: 135-138.

Johnston, J. R. 1985. "How Personality Attributes Structure Interpersonal Relations." In *Status, Rewards, and Influence,* edited by J.Berger and M. Zelditch. San Francisco: Jossey-Bass.

Jorgenson, D. 1975. "Economic Threat and Authoritarianism in Television Programs: 1950-1974." *Psychological Reports* 35: 1153-1154.

Kaduson, H. G., and C. E. Schaefer. 1991. "Concurrent Validity of the Creative Personality Scale of the Adjective Check List." *Psychological Reports* 69: 601-602.

Kalick, S. 1988. "Physical Attractiveness as a Status Cue." *Journal of Experimental Social Psychology* 24: 469-489.

Kaplan, R. 1978. "Is Beauty Talent? Sex Interaction in the Attractiveness Halo Effect." *Sex Roles* 4: 195-203.

Karlins, M., T. Coffman, and G. Walters. 1969. "On the Fading of Social Stereotypes: Studies in Three Generations of College Students." *Journal of Personality and Social Psychology* 13: 1-16.

Kasof, J. 1992. "Social Determinants of Creativity." Unpublished doctoral dissertation, Harvard University.

————. 1993. "Sex Bias in the Naming of Stimulus Persons." *Psychological Bulletin* 113:140-163.

————. In press. "Explaining Creativity: The Attributional Perspective." *Creativity Research Journal.*

Katz, A. and L. Giacommelli. 1982. "The Subjective Nature of Creativity Judgments." *Bulletin of the Psychonomic Society* 20: 17-20.

Kite, M. and B. Johnson. 1988. "Attitudes Toward Older and Younger Adults: A Meta-analysis." *Psychology and Aging* 3: 233-244.

Kohn, M. L. and C. Schooler. 1969. "Class, Occupation, and Orientation." *American Sociological Review* 34: 659-678.

————. 1982. "Job Conditions and Personality: A Longitudinal Assessment of their Reciprocal Effects." *American Journal of Sociology* 87: 1257-1286.

Kohn, M. L., A. Naoi, C. Schoenbach, C. Schooler, and K. M. Slomczynski. 1990. "Position in the Class Structure and Psychological Functioning in the United States, Japan, and Poland." *American Journal of Sociology* 95: 964-1008.

Kollock, P., P. Blumstein, and P. Schwartz. 1985. "Sex and Power in Interaction: Conversational Privileges and Duties." *American Sociological Review* 50: 34-46.

Krueger, J. and M. Rothbart. 1988. "Use of Categorical and Individuating Information in Making Inferences about Personality." *Journal of Personality and Social Psychology* 55: 187-195.

Kruglanski, A. and T. Freund. 1983. "The Freezing and Unfreezing of Lay-inferences: Effects on Impressional Primacy, Ethnic Stereotyping, and Numerical Anchoring." *Journal of Experimental Social Psychology* 19: 448-468.

Kuhn, T. S. 1957. *The Copernican Revolution.* Cambridge, MA: Harvard University Press.

Landy, D. and H. Sigall. 1974. "Beauty is Talent: Task Evaluation as a Function of the Performer's Physical Attractiveness." *Journal of Personality and Social Psychology* 29: 299-304.

Lang, G. E. and K. Lang. 1988. "Recognition and Renown: The Survival of Artistic Reputation." *American Journal of Sociology* 94: 79-109.

Leffler, A., D. Gillespie, and J. Conaty. 1982. "The Effects of Status Differentiation on Nonverbal Behavior." *Social Psychology Quarterly* 45: 153-161.

Lenney, E., L. Mitchell, and C. Browning. 1983. "The Effect of Clear Evaluation Criteria on Sex Bias in Judgments of Performance." *Psychology of Women Quarterly* 7: 313-328.

Levenson, H., B. Burford, B. Bonno, and L. Davis. 1975. "Are Women Still Prejudiced Against Women? A Replication and Extension of Goldberg's Study." *Journal of Psychology* 89: 67-71.

Li, J. and H. Gardner. 1993. "How Domains Constrain Creativity." *American Behavioral Scientist* 37: 94-101.

Lipton, J. and A. Hershaft. 1984. "'Girl,' 'woman,' 'guy,' 'man': The Effects of Sexist Labeling." *Sex Roles* 10: 183-194.

Locksley, A., E. Borgida, N. Brekke, and C. Hepburn. 1980. "Sex Stereotypes and Social Judgment." *Journal of Personality and Social Psychology* 39: 821-831.

Locksley, A., C. Hepburn, and V. Ortiz. 1982. "Social Stereotypes and Judgments of Individuals: An Instance of the Base-rate Fallacy." *Journal of Experimental Social Psychology* 18: 23-42.

Long, J., P. Allison, and R. McGinnis. 1979. "Entrance into the Academic Career." *American Sociological Review* 44: 816-830.

Lord, C., M. Lepper, and D. Mackie. 1984. "Attitude Prototypes as Determinants of Attitude-behavior Consistency." *Journal of Personality and Social Psychology* 46: 1254-1266.

MacKinnon, D. 1964. *The Study of Lives.* New York: Atherton Press.

Madigan, C. and A. Elwood. 1983. *Brainstorms and Thunderbolts: How Creative Genius Works.* New York: Macmillan.

Maquet, J. 1986. *The Aesthetic Experience: An Anthropologist Looks at the Visual Arts.* New Haven: Yale University Press.

Markovsky, B., L. Smith, and J. Berger. 1984. "Do Status Interventions Persist?" *American Sociological Review* 49: 373-382.

Martindale, C. 1975. *Romantic Progression.* Washington, DC: Hemisphere.

Maruyama, G. and N. Miller. 1980. "Physical Attractiveness, Race, and Essay Evaluation." *Personality and Social Psychology Bulletin* 6: 384-490.

Mazur, A., E. Rosa, M. Faupel, J. Heller, R. Leen, and B. Thurman. 1980. "Physiological Aspects of Communication Via Mutual Eye Gaze." *American Journal of Sociology* 86: 50-74.

Mead, M. 1935. *Sex and Temperament in Three Primitive Societies.* New York: Morrow.

Michael, W., B. Rosenthal, and M. De Camp. 1949. "An Experimental Investigation of Prestige-suggestion for two Types of Literary Material." *Journal of Psychology* 28: 303-323.

Miller, A. 1970. "Role of Physical Attractiveness in Impression Formation." *Psychonomic Science* 19: 241-243.

Miller, N., G. Maruyama, R. J. Beaber, and K. Valone. 1976. "Speed of Speech and Persuasion." *Journal of Personality and Social Psychology* 34: 615-624.

Mischel, H. 1974. "Sex Bias in the Evaluation of Professional Achievements." *Journal of Educational Psychology* 66: 157-166.

Mohr, P. 1986. "Demeanor, Status Cue or Performance?" *Social Psychology Quarterly* 49:228-236.

Moore, H. 1921. "The Comparative Influence of Majority and Expert Opinion." *American Journal of Psychology* 32: 16-20.

Mullen, B., C. Johnson., and E. Salas. 1991. "Productivity Loss in Brainstorming Groups: A Meta-analytic Integration." *Basic and Applied Social Psychology* 12: 3-24.

Murphy, M. and D. Hellkamp. 1976. "Attractiveness and Personality Warmth: Evaluations of Paintings Rated by College Men and Women." *Perceptual and Motor Skills* 43: 1163-1166.

Nemeth, C. and J. Wachtler. 1974. "Creating the Perceptions of Consistency and Confidence: A Necessary Condition of Minority Influence." *Sociometry* 37: 529-540.

Noel, R. and M. Allen. 1976. "Sex and Ethnic Bias in the Evaluation of Student Editorials." *Journal of Psychology* 94: 53-58.

Ochse, R. 1990. *Before the Gates of Excellence: The Determinants of Creative Genius.* Cambridge: Cambridge University Press.

Osborn, A. F. 1957. *Applied Imagination.* New York: Scribner.

Palmore, E. and K. Manton. 1974. "Modernization and Status of Aged: International Correlations." *Journal of Gerontology* 29: 205-210.

Paludi, M. and W. Bauer. 1983. "Goldberg Revisited: What's in an Author's Name." *Sex Roles* 9: 387-390.

Paludi, M. and L. Strayer. 1985. "What's in an Author's Name? Differential Evaluations of Performance as a Function of Author's Name." *Sex Roles* 12: 353-361.

Panek, P., R. Deitchman, J. Burkholder, T. Speroff, and R. Haude. 1976. "Evaluation of Feminine Professional Competence as a Function of Level of Accomplishment." *Psychological Reports* 38: 875-880.

Patterson, O. 1982. *Slavery and Social Death.* Cambridge, MA: Harvard University Press.

Paulus, P. B. and M. T. Dzindolet. 1993. "Social Influence Processes in Group Brainstorming." *Journal of Personality and Social Psychology* 64: 575-586.

Paulus, P. B., M. T. Dzindolet, G. Poletes, and L. M. Camacho. 1993. "Perception of Performance in Group Brainstorming: The Illusion of Group Productivity." *Journal of Personality and Social Psychology* 19: 78-89.

Peck, T. 1978. "When Women Evaluate Women, Nothing Succeeds Like Success: The Differential Effects of Status upon Evaluations of Male and Female Professional Ability." *Sex Roles* 4: 205-213.

Pheterson, G., S. Kiesler, and P. Goldberg. 1971. "Evaluation of the Performance of Women as a Function of their Sex, Achievement, and Personal History." *Journal of Personality and Social Psychology* 19: 114-118.

Pleban, R. and A. Tesser. 1981. "The Effects of Relevance and Quality of Another's Performance of Interpersonal Closeness." *Social Psychology Quarterly* 44: 278-285.

Puckett, J., R. Petty, J. Cacioppo, and D. Fischer. 1983. "The Relative Impact of Age and Attractiveness Stereotypes on Persuasion." *Journal of Gerontology* 38: 340-343.

Rickard, K. M. 1990. "The Effect of Feminist Identity Level on Gender Prejudice toward Artists' Illustrations." *Journal of Research in Personality* 24: 145-162.

Ridgeway, C. L. 1982. "Status in Groups: The Importance of Motivation." *American Sociological Review* 47: 76-88.

————. 1987. "Nonverbal Behavior, Dominance, and the Basis of Status in Task Groups." *American Sociological Review* 52: 683-694.

————. (Ed.). 1992. *Gender, Interaction, and Inequality.* New York: Springer-Verlag.

Ridgeway, C. and D. Diekema. 1989. "Dominance and Collective Hierarchy Formation in Male and Female Task Groups." *American Sociological Review* 54: 979-93.

Ridgeway, C., J. Berger, and L. Smith. 1985. "Nonverbal Cues and Status: An Expectation States Approach." *American Journal of Sociology* 90: 955-978.

Ridgeway, C. and J. Berger. 1986. "Expectations, Legitimation, and Dominance Behavior in Task Groups." *American Sociological Review* 51: 603-617.

Rosa, E. and A. Mazur. 1979. "Incipient Status in Small Groups." *Social Forces* 58: 18-37.

Rosen, B. and T. Jerdee. 1976. "The Influence of Age Stereotypes on Managerial Decisions." *Journal of Applied Psychology* 61: 428-432.

Rossman, B. B. and H. F. Gollub. 1975. "Comparison of Social Judgments of Creativity and Intelligence." *Journal of Personality and Social Psychology* 31: 271-281.

Ruback, R. B., J. M. Dabbs, and C. H. Hopper. 1984. "The Process of Brainstorming: An Analysis with Individual and Group Vocal Parameters." *Journal of Personality and Social Psychology* 47: 558-567.

Runco, M. A. and M. D. Bahleda. 1986. "Implicit Theories of Artistic, Scientific, and Everyday Creativity." *Journal of Creative Behavior* 20: 93-98.

Runco, M. A. and R. E. Charles. 1993. "Judgments of Originality and Appropriateness As Predictors of Creativity." *Personality and Individual Differences* 15: 537-546.

Runco, M. A. and W. R. Smith. 1992. "Interpersonal and Intrapersonal Evaluations of Creative Ideas." *Personality and Individual Differences* 13: 295-302.

Runco, M. A. and L. Vega. 1990. "Evaluating the Creativity of Children's Ideas." *Journal of Social Behavior and Personality* 5: 439-452.

Sachdev, I. and R. Y. Bourhis. 1987. "Status Differentials and Intergroup Behaviour." *European Journal of Social Psychology* 17: 277-293.

————. 1991. "Power and Status Differentials in Minority and Majority Group Relations." *European Journal of Social Psychology* 21: 1-24.

Sales, S. 1973. "Threat as a Factor in Authoritarianism: An Analysis of Archival Data." *Journal of Personality and Social Psychology* 28: 44-57.

Sev'er, A. 1989. "Simultaneous Effects of Status and Task Cues: Combining, Eliminating or Buffering?" *Social Psychology Quarterly* 52: 327-335.

Sherif, M. 1935. "An Experimental Study of Stereotypes." *Journal of Abnormal and Social Psychology* 30: 371-375.

Shrum, W. 1991. "Critics and Publics: Cultural Mediation in Highbrow and Popular Performing Arts." *American Journal of Sociology* 97: 347-375.

Simon, B., B. Glassner-Bayerl, and I. Stratenwerth. 1991. "Stereotyping and Self-stereotyping in a Natural Intergroup Context: The Case of Heterosexual and Homosexual Men." *Social Psychology Quarterly* 54: 252-266.

Simonton, D. K. 1984. *Genius, Creativity, and Leadership.* Cambridge, MA: Harvard University Press.

————. 1988. "Creativity, Leadership, and Chance." Pp. 386-426 in *The Nature of Creativity,* edited by R. J. Sternberg. Cambridge: Cambridge University Press.

————. 1992a. "Leaders of American Psychology, 1879-1967: Career Development, Creative Output, and Professional Achievement." *Journal of Personality and Social Psychology* 62: 5-17.

————. 1992b. "Gender and Genius in Japan: Feminine Eminence in Masculine Culture." *Sex Roles* 27: 101-119.

Sinha, A. and O. Upadhyaya. 1960. "Change and Persistence in the Stereotypes of University Students Toward Different Ethnic Groups During the Sino-Indian Border Dispute." *Journal of Social Psychology* 52: 31-39.

Snyder, C. R., M. Lassegard, and C. Ford. 1986. "Distancing After Group Success and Failure: Basking in Reflected Glory and Cutting off Reflected Failure." *Journal of Personality and Social Psychology* 51: 382-388.

Sorokin, P. A. 1937-1941. *Social and Cultural Dynamics* (4 volumes). New York: American Books.

Starer, R. and F. Denmark. 1974. "Discrimination Against Aspiring Women." *International Journal of Group Tensions* 4: 65-70.

Sternberg, R. J. 1985. "Implicit Theories of Intelligence, Creativity, and Wisdom." *Journal of Personality and Social Psychology* 49: 607-627.

Swim, J., E. Borgida, G. Maruyama, and D. Myers. 1989. "Joan McKay versus John McKay: Do Gender Stereotypes Bias Evaluation?" *Psychological Bulletin* 105: 409-29.

Tanner, L. 1977. "Sex Bias in Children's Response to Literature." *Language Arts* 54: 48-50.

Tetlock, P. E. 1988. "Integrative Complexity of Text and of Reader as Moderators of Aesthetic Evaluations." *Poetics* 17: 357-366.

Toder, N. L. 1980. "The Effect of the Sexual Composition of a Group on Discrimination Against Women and Sex-role Attitudes." *Psychology of Women Quarterly* 5: 292-310.

Top, T. J. 1991. "Sex Bias in the Evaluation of Performance in the Scientific, Artistic, and Literary Professions: A Review." *Sex Roles* 24: 73-106.

Torrance, E. P. 1963. "Changing Reactions of Preadolescent Girls to Tasks Requiring Creative Scientific Thinking." *Journal of Genetic Psychology* 102: 217-223.

Tosi, H. and S. Einbender. 1985. "The Effects of the Type and Amount of Information in Sex Discrimination Research: A Meta-analysis." *Academy of Management Journal* 28: 712-723.

Tuchman, G. and N. Fortin. 1984. "Fame and Misfortune: Edging Women out of the Great Literary Tradition." *American Journal of Sociology* 90: 72-96.

Valasek, D., B. Avolio, and L. Forbringer. 1979. "Effect of Sex-stereotyping in Evaluating Males in a Traditional Female Role." *Psychological Reports* 44: 1196-1198.

Van Oudenhoven, J., F. Siero, and J. Withag. 1984. "La Langue Officielle et l'Evaluationde Redactions Composees par des Eleves de Milieux Socio-culturels Differents." *Psychologie Francaise* 29: 204-208.

Wagner, D. G. and J. Berger. 1982. "Paths of Relevance and the Induction of Status-task Expectancies: A Research Note." *Social Forces* 61: 575-586.

Wallace, R. 1969. *The World of Van Gogh: 1853-1890.* New York: Time-Life Books.

Wallston, B. and V. O'Leary. 1981. "Sex makes a Difference: Differential Perceptions of Women and Men." *Review of Personality and Social Psychology* 2: 9-41.

Walsh, R. and C. Connor. 1979. "Old Men and Young Women: How Objectively are Their Skills Assessed?" *Journal of Gerontology* 34: 561-568.

Ward, C. 1978. "Methodological Problems in Attitude Measurement: Sex Roles, Social Approval and the Bogus Pipeline." *Representative Research in Social Psychology* 9: 64-68.

————. 1979. "Differential Evaluation of Male and Female Expertise: Prejudice Against Women?" *British Journal of Social and Clinical Psychology* 18: 65-69.

————. 1981. "Prejudice Against Women: Who, When, and Why?" *Sex Roles* 7: 163-171.

Ward, D. and J. Balswick. 1978. "Strong Men and Virtuous Women: A Content Analysis of Sex Role Stereotypes." *Pacific Sociological Review* 21: 45-53.

Watt, I. 1957. *The Rise of the Novel.* Berkeley: University of California Press.

Webster, M. 1977. "Equating Characteristics and Social Interaction: Two Experiments." *Sociometry* 40: 41-50.

Webster, M. and J. E. Driskell. 1978. "Status Generalization: A Review and Some New Data." *American Sociological Association* 43: 220-236.

————. 1983. "Beauty as Status." *American Journal of Sociology* 89: 140-165.

Webster, M. and M. Foschi (Eds.). 1988. *Status Generalization: New Theory and Research.* Stanford, CA: Stanford University Press.

Wehner, L., M. Csikszentmihalyi, and I. Maguari-Beck. 1991. "Current Approaches used in Studying Creativity: An Exploratory Investigation." *Creativity Research Journal* 4: 261-271.

Weisberg, R. W. 1986. *Creativity: Genius and Other Myths.* New York: Freeman.

White, H. C. 1993. *Careers and Creativity: Social Forces in the Arts.* Boulder, CO: Westview Press.

Willard, D. and F. L. Strodtbeck. 1972. "Latency of Verbal Response and Participation in Small Groups." *Sociometry* 35: 161-175.

Wood, W. and S. J. Karten. 1986. "Sex Differences in Interaction Style as a Product of Perceived Sex Differences in Competence." *Journal of Personality and Social Science* 50: 341-347.

Zuckerman, H. A. 1977. *Scientific Elite: Nobel Laureates in the U.S.* New York: Free Press.

Zuckerman, H. A. and R. K. Merton. 1971. "Patterns of Evaluation in Science: Institutionalization, Structure and Function of the Referee System." *Minerva* 9: 66-100.

ROLE AS RESOURCE IN A PRISON UPRISING

Peter L. Callero

ABSTRACT

In 1988, inmates at the Oregon Women's Correctional Center staged a sit-down resistance in protest of conditions within the prison. The demonstration was effectively quelled by authorities in a matter of hours. This small uprising is used as a case study to illustrate the value of conceptualizing roles as resources. It is argued that the resource perspective transcends the conventional division between interactionist and structural approaches to role theory by offering a framework that recognizes roles as elements in the production of both agency and structure. Important conceptual distinctions between variation in role type and variation in role use are demonstrated using personal interviews and historical documents pertaining to the prison uprising.

Dating back to at least the work of Linton (1936), the role concept has been used by sociologists to explain and describe a variety of sociological events. At times the approach has been one in which the role concept has been the

Advances in Group Processes, Volume 12, pages 221-247.
Copyright © 1995 by JAI Press Inc.
All rights of reproduction in any form reserved.
ISBN: 1-55938-872-2

central organizing component of analyses. This was the case, for example, in Gross, McEachern, and Mason's (1957) classic study of school administrators and Cottrell's (1942) work on age and gender. Since this time, however, the role concept has typically been used in a more limited manner, either as a bridge between micro and macro levels of analysis or as a basis for explaining variation in individual behavior (cf. Stryker and Statham 1985).

In this paper I will illustrate the value of returning to a more extensive role analysis that aims to link social structure and agency. I want to stress, however, that the strategy of this analysis departs radically from the traditional functional approach advocated by theorists such as Merton (1957) and Goode (1960). Instead of viewing roles as expectations associated with positions in social structures, roles will be conceptualized as resources that vary both in terms of how they are structured and how they are used.

The resource approach to role theory combines elements from the work of interactionists (Hewitt 1989; Turner 1962), ethnomethodologists (Hilbert 1981), conflict theorists (Gerhardt 1980), and network analysts (White, Boorman, and Breiger 1976). The most formal development of the perspective, however, appears in a paper by Baker and Faulkner (1991), as well as a very recent paper of my own (Callero 1994).

The empirical example I will be working with in this chapter is a small "uprising" that occurred at a women's prison in the state of Oregon. I hope to use this particular event to illustrate the utility of the resource perspective and the value of research that takes roles as central to both social organization and social change. The primary intent of this analysis, then, is not to explain the causes of prison uprisings or even the emergence of social movements in general. Instead, my goal here is to demonstrate the unique insight offered by a resource based role analysis.

When using the resource perspective as a theoretical framework, certain fundamental research questions become evident. I will touch upon each of these in this chapter: First, how can both social structure and human agency be simultaneously maintained through social roles? Second, in what different ways do roles serve as resources? Third, what are the different types of role resources, and fourth, how do these differences condition the various uses of roles? Before exploring these questions, however, I will need to begin with a description of the event to be examined.

THE PRISON UPRISING[1]

October 1, 1988 was an unusually warm Saturday in Salem, Oregon, a city of approximately 100,000. By 1:00 pm the temperature had reached a record 90 degrees downtown along the Willamette River. Two miles away at the Oregon Women's Correctional Center (OWCC) about 70 women inmates had entered the prison yard for their usual afternoon recreational period. If it weren't for a 15-foot high chain-link fence topped by coils of razor wire, the grounds could easily be mistaken for a simple office complex. Indeed, the one-story brick buildings of the women's prison are literally overshadowed by the massive stone walls and armed turrets of the neighboring men's prison. On this day, however, the women's institution was about to generate its own focus of public attention.

About an hour into the standard three-and-one-half hour recreational period the rumor of a possible inmate disturbance was circulating among correctional officers. Still, with the exception of a group of about ten women congregating near the edge of the yard, nothing appeared unusual or out of the ordinary to the seven officers on surveillance.

At about 3:30 pm, however, at a downtown office building, a dispatcher for the local newspaper received an anonymous tip that an inmate protest was underway at the women's prison. In routine fashion a reporter and photographer were sent to investigate the lead. They arrived in the parking lot of the medium security prison within minutes. As they approached the wire fence surrounding the recreational yard they quickly scanned the small grass covered enclosure. There was no visible evidence of a protest or demonstration. Thirty to forty women dressed in the standard issue blue denim trousers and navy blue tee shirts were milling about. A group of about ten women were huddled together around a picnic table. To the reporter, the tip appeared to be a hoax. Within minutes, however, a middle-aged man entered the parking lot in an unmarked car, jumped out, and headed straight for the reporters, who had now been joined by a visitor to the prison. All three were ordered to immediately leave the grounds under threat of arrest.

The confrontation in the parking lot was over in a matter of minutes. After the newspaper staff members and visitor were escorted to the street, Oregon State Patrol officers were posted at the prison entrance. All subsequent requests to enter the grounds were denied.

It was now 4:00 pm and inside the prison a protest action was in fact slowly being constructed. The superintendent's office received a call from the *Oregonian*, the state's largest newspaper. They too had received a tip that a protest was underway and were seeking verification. No comment was provided. Meanwhile, correctional officers in the yard were instructed to record the names of any inmates participating in unusual behavior. A video taped recording of inmate activity was also initiated. The group of women at one end of the yard was now growing in size. Most were sitting together in a circle formation on the grass, animated in excited discussion. Cheers and applause would erupt as others from the yard entered the circle.

At 4:15 pm, a 31-year-old inmate emerged from the group and approached a correctional officer. She held in her hand a four page declaration of protest. Neatly hand printed on lined notebook paper was the following statement:[2]

Let it be known to Staff and Administration of Oregon Women Correctional Center that on todays date October 1, 1988 a quiet, peaceful sitdown demonstration has been called together by the Inmate Residents of Oregon Womens Correction Center. This demonstration has been called in the inmate recreation yard and will disperse only upon agreement by and with R. H. Sheidler (Superintendent), Michael Frankie (Director of Corrections) and Mr. Neil Goldschmidt (Governor of the State of Oregon) to take our requests into "Serious" consideration with results being beneficial to us inmates. We would ask that those taking part in this demonstration not be subjected to any violence and or any serious Disciplinary Actions. Our requests are not exagerated or unreasonable. Living conditions inside O.W.C.C. have reached a point where if not for this peaceful demonstration the likelihood of a full scale riot would be realized. The issues involved are of Safety and Sanitary nature, Education Programs, Rehabilitation Programs, Housing situation, lack of decent Medical and Dental care. Our requests are listed on the attached pages with cases in point to clarify our reasons for these requests. Let it also be known that there are inmates who have sworn to a Hunger Strike until these issues are met and resolved. Requests:

1. That a copy of the statement and requests be made available to the Press.
2. That the designated spokespersons for the inmates be allowed to speak with and to the Press as well as Mr.'s Schiedler, Frankie and Goldschmidt.
3. That during the course of this demonstration toilet facilities be made available to the inmates.
4. We would ask that better and faster medical attention be provided by the Institution. Case in point: It takes several weeks for Interview Requests to be answered. There is no Follow-up care given.
5. We would ask that better and faster Dental care be provided. Case in point: At this time it takes several weeks to see the dentist even in Emergency situation. One inmate had to have teeth pulled due to an infection that was left to fester all of Memorial Day weekend. That inmate has not recieved a partial plate to this date even though inmate can and will pay for the partial plate.
6. We would be asked to be given back our Higher Education Program. This Fall the residents of O.W.C.C. can no longer attend Higher Education courses offered at O.S.P. unless we are already in possession of an Associates of Arts degree. At this time we are only be offered 5 different 1st sequence course here at O.W.C.C. In this act of education denial, [illegible] & psych. care is also denied. These are also available in a higher education program.
7. We would ask that an Honor Living situation be provided for those inmates doing long terms, working and or going to school. We would also ask that long term inmates be given the right to choose congenial room-mates.

8. We would ask that we be given disinfectant to use in our rooms and the showers. Case point: At this time there are only 2 showers to be shared between the entire General Population and we aren't given any disinfectant to use in between showers.

9. We would ask for the return of our Club Sales, which were taken due to a ruling on inmates giving monies or selling to inmates. Case Points: At this time we would be allowed to purchase photos from the institutions with the money going into the inmate welfare account thereby the money is still going to be going from inmate to inmate.

10. We would ask that staff and administration be more aware and considerate of inmates safety and health. Case Point: At this time there is construction going on inside the institution. Filter masks were issued to staff and not to inmates to protect staff from the chalk and concrete dust. There is a writ currently awaiting appeal for air quality.

11. We would ask that the same sort of drainage be done for the existing showers. Case Point: At this time there is continuously several inches of standing water on the floor and in the shower. This being unsafe and unsanitary.

12. We would ask that Administration consider the taking away of several things in one years time. Case Point: Christmas Boxs, Incentive Programs, Personal Plants, Closet Curtains, Stuffed animals and materials to make stuffed animals for our family and friends, club sales, and now rumor has it our visiting is to become Security Visit only. Also, clubs at O.W.C.C. and O.S.P. are no longer allowed to share their annuals or Pow-wow. We ask that the Administration reconsider the taking away of these priveleges.

13. We would ask that during the course of this demonstration the staff at O.W.C.C. be responsible for providing us inmates with food stuff should we need it.

14. We would ask that something be done to immediately relieve the severe overcrowding here at O.W.C.C. Case Point at this time there are several single bunk rooms in west (over)....[3]

Following a brief set of instructions, the correctional officer delivered the statement to the commanding officer who in turn took it to the office of the superintendent. It was now formalized: a sit-down strike was about to take place and both sides were poised for the inevitable confrontation.

At exactly 4:30 pm the regularly scheduled "yard call" began. It is at this moment that all inmates in the yard are required to return to their cells in orderly fashion for a head count. For years prior to this day the procedure was followed without deviation. Today, however, 28 women sat in a circle and ignored the official command to return to their cells. While the 40 women who chose not to protest were escorted back inside the prison, the defiant protesters joined in singing the chorus of John Lennon's ballad "Give Peace a Chance." With a slight alteration of the original lyrics, they sang "All we are saying, is let us be heard."

Inside the prison, the correctional management staff were also beginning to mobilize. The top prison staff who had been off duty on this day, arrived and were now in consultation. They immediately called for State Police assistance in covering the perimeter of the complex. The neighboring maximum security men's prison was also alerted, agreeing immediately to provide perimeter surveillance from guard posts on the penitentiary wall. Five top officials from both the Department of Corrections and the two men's prisons in the area arrived to help coordinate a response.

By 5:00 pm a planned response to the sit-down was underway. The west residential wing of the prison was vacated. The non-protesting prisoners were moved into cells with other inmates and the west wing cells were completely emptied of books, clothes, personal items, linens and all furniture, except for the bunks and a desk. Approximately 25 off-duty correctional officers were called at home and ordered to report to the prison to assist with the "state of emergency." The protesters were still in the yard, sitting in a circle, when the security manager made the following command over a hand held bull horn: "You are involved in an illegal activity, report to your cells immediately! If you do not, legal action will be taken!" The order was ignored.

It was now 5:30 pm and the protesters had been alone in the locked yard for an hour, anxiously anticipating an official response. Another note was written and placed in a can at the locked entrance by the same women who had delivered the original declaration of protest. She was ordered away from the gate and the message was retrieved by a correctional officer and delivered to prison officials huddled in consultation inside the superintendent's office. The note listed the names of protesters who were in need of their scheduled medication. The request for medication was ignored.

By 6:00 pm the west wing cells had been completely cleared and prison officials had agreed upon plans for an intervention. First, an armored vehicle containing three officers arrived from the men's prison and began patrolling a small area on the outside of the fence facing the protesters. Next, a "tactical emergency response team" (TERT) composed of seven specially trained officers from the men's prison assembled inside the prison dayroom. All were in riot gear and outfitted with helmets and batons. Three members of the team were armed with shotguns.

At 6:20 pm an order was given to end the protest. The response team entered the yard and marched in a deliberate wedge formation toward the group of protesters. In the middle of the wedge stood the security manager with bull horn in hand. The protesters, still sitting in a circle with their backs to each other, reacted quickly by linking arms in a display of solidarity and defiance. At the same time, inmates who were watching the yard from their cell windows in the east wing reacted in fear and anger at the sight of the armed group heading toward the circle of women. In a demonstration of support for the protesters, inmates inside the prison began to pound on windows and walls while at the same time screaming and yelling their objections to the TERT force.

At a point about halfway between the yard gate and the group of protesters, the TERT squad came to a halt. The security manager shouted through her bullhorn, "You are involved in an illegal activity! Lie face down on the ground!" The protesters refused to comply. The same order was given twice more and was ignored both times. Once again, the response team began its deliberate march toward the group where many of the protesters were now fearing for their lives.

As the response team approached the group, the three officers carrying shotguns quickly maneuvered into standing positions on the perimeter of the circle and aimed their weapons at the women. In coordinated action, the other officers pulled the protesters apart, forced each one to the ground face first, and cuffed their hands behind their backs with strong plastic restraints. One by one each protester was then escorted back into the prison dayroom where they were handed over to a team of waiting officers (both male and female) who ordered the removal of all clothing for the purpose of a full body "skin search."

After the body search, the protesters were handed robes and were escorted to the empty cells in the west wing. By 6:45 pm the yard had been completely cleared. The sit-down strike had been effectively ended with only minor physical injuries to two inmates. One twenty-five-year old inmate complained of a wrenched shoulder, and another thirty-two-year old inmate who had experienced a seizure, was transported to a local hospital for observation.

Although the yard was now quiet, the banging and yelling that began in the east wing was still occurring, and in fact had spread to the west wing where the protesters were housed. Correctional officers tried aggressively to quiet the inmates without success. For the next three hours the women's prison was filled with protest songs, shouts of anger, and the banging of hands and other objects on the metal cell doors. It was 10:00 pm before the disturbance had completely subsided.

STRUCTURE AND AGENCY

Clearly, the sit-down protest described here can be understood from numerous sociological perspectives. I would, however, like to take it as a case study for illustrating key elements of a particular type of role analysis. As I noted in the introduction, the role analysis that I am advocating begins by conceptualizing roles as social resources. The implications of this premise are many, but perhaps the most important and general contribution of the resource perspective is its promise for articulating the relationship between structure and agency. This duality, which in many ways is at the heart of sociological analysis, has historically defined the division within role theory between structural and interactionist approaches. While the structuralists are said to ignore the importance of agency and social change, the interactionists are critiqued for their naive articulation of power, tradition, and other limits to action. The resource perspective transcends this division at both a conceptual and analytical level by offering a framework that recognizes roles as elements in the production of both agency and structure.

Schemas and Resources

The emerging consensus among sociologists is that society consists of both powerful, determining structures *and* actors that demonstrate a degree of efficacy, freedom, and creative independence. A conceptualization of structure that recognizes this essential duality has been developed by Giddens (1984) and elaborated more recently by Sewell (1992). According to Giddens, structure is "carried" in reproduced practices and relationships and, therefore, depends highly upon interaction for its maintenance. More specifically, it is argued that structure exists in two distinct but mutually dependent dimensions, what Sewell refers to as "schemas" and "resources." Schemas are the taken for granted rules and cultural assumptions that serve as principles of action, while resources are human qualities and nonhuman objects that serve as a source of power. For example, in a typical prison setting, schemas are evident in the assumptions that structure the relation among inmates (e.g., doing long time is worthy of respect), among prison employees (e.g., one follows orders of a superior officer), and between inmates and prison employees (e.g., one should not trust the other). Resources are evident in both the nonhuman material conditions of the prison (e.g., weapons, prison walls) as well as the human qualities of deference, fear, and/or authority that are granted to judges, lawyers, and correctional officers vis-à-vis convicts.

While schemas and resources combine to form structure, there is an important difference between these two dimensions. Schemas are said to have a "virtual" existence in that they are not reducible to any particular location in space in time. The schemas that serve to distinguish the relations among inmates and guards, for example, are generalizable procedures that can be put into practice in a range of different settings. Resources, on the other hand, are conceptualized as having an "actual" existence, meaning they are observable in time and space. The prison walls, the weapons, the actual following of orders, all have an actual existence and serve as a source of power. Nevertheless, it should be reemphasized that the two dimensions cannot be maintained independently. The actualization or enactment of cultural schemas produces the observable human resources and the use of resources validates the reality of the schemas. Resources are the product of schemas and schemas are generated from resources.

The schemas and resources that combined to structure the moment of the sit-down protest are multi-layered and complex. At the most general level, we can assume that the incident was structured by the same schemas that overlay all ordinary interactions among all people in society. Similarly, we can also assume that there are resources that are not particular to this interaction but are available to all whether inside or outside the prison. What is important analytically, however, are the schemas and resources that were salient and/ or unique to the uprising. In other words, what specific schemas and resources structured the sit-down strike?

Without the framework of a more specific middle-range theory, this is a difficult question to answer. I believe, however, that the role concept can play an important part in filling this conceptual void. If we view roles as elements of structure we can see more clearly how schemas and resources reflect both structure and agency.

Role as Social Structure

Consistent with Sewell's elaboration of Giddens, I want to argue that a role represents particular *sets* of schemas and resources. In other words, roles must be seen as both schema and resource.[4] As a schema, a role has a *virtual* reality in the sense that it exists at a transcendent level, available for use in different intersections of space and time. A role, therefore, is not dependent upon a particular status or position and is not simply a set of expectations. It is a generalizable set of meanings or principles (Callero 1994). It is at this level that we can say a role has a cognitive dimension (Callero 1991), and serves as an image (Schwalbe 1987) or gestalt (Turner 1962) that directly or indirectly guides action. When employed as a resource, a role also has an *actual* existence, and is the basis of power. At this level a role is observable as it is enacted in particular encounters and becomes a real object of interaction.

It is important to emphasize once again that a role is simultaneously both schema and resource. The particular aspects of a role as a schema are linked to particular features of a role as resource. This distinction can be understood more concretely for the case of social roles when we recognize that roles differ in terms of *type* (e.g., inmate vs. guard) and in terms of *use*. The way in which a role is *used* depends upon the *type* of role it is. Similarly, the reality of a role's *type* is experienced, known, and affirmed in its *use*.

I want to suggest that the dual nature of roles corresponds to the dual nature of social structure and that linking role and social structure in this way has certain advantages over more familiar conceptualizations of role. For example, it avoids the overly deterministic approach to action characteristic of traditional structural approaches (e.g., Linton 1936), while at the same time it offers a conceptual link to a dynamic theory of social structure, a quality absent from most interactionist approaches (e.g., Turner 1962). As a consequence, the role as resource approach is better able to articulate processes of social change as they emerge from collective action. Other theories of role that attempt to link agency and structure often focus on the explanation of individual action within a structural context and do not address the changes in social structure that result from using roles. This is true, for example, of Stryker's (1980) "social structural version" of symbolic interactionism, where questions regarding the performance of a role, the commitment to a role, and the relationship between role and identity, take precedence over questions of structural change. Moreover, even though Stryker (1980, pp. 57-59) recognizes the fluid and dynamic aspects of role-making, he nevertheless defines roles in terms of normative expectations. Defining role in this way establishes an overly deterministic conceptualization of structure and severely limits the conceptualization of agency.[5] The role as resource perspective offers a more encompassing framework where the mutual interdependence of agency and structure is articulated in a manner that avoids the problematic dualisms of more traditional conceptualizations of role. The essential insight of the resource perspective is straightforward: roles are used as tools for creative action, agency and change, and the agency experienced from using roles is limited and structured by the meaning or type of role. In addition, the cultural variation in role type (virtual existence) is reproduced through role use (actual existence). The uprising at the Oregon women's prison can provide a case study for illustrating some of the implications of this basic idea.

Variation in Role Type

The idea that there are different types of roles is not new. Sociologists have made traditional distinctions, for example, between expressive and instrumental roles (Parsons and Shils 1951), task and status roles (Bales 1958), and roles associated with ascribed and achieved statuses (Davis 1949). But when roles are conceptualized as elements of social structure linked to particular sets of schemas that generalize across time and space, a whole new and different array of variation in role types becomes evident. Four categories of variation are particularly important. These dimensions combine to structure the unique sets of meanings associated with different roles.

Cultural Endorsement

At the most obvious level, we can see that roles differ by name and by the meanings associated with the names. During the protest at the Oregon Women's Correctional Center, for example, one could identify very generally, the roles of inmate, protester, prison guard, tactical team member, prison administrator, and reporter, among others. These roles are for the most part formally defined by bureaucratic rules and also share a general acceptance by the wider community as legitimate cultural objects. In other words, most of us outside of the prison could use these roles to make sense of the events of the protest. Indeed, when the reporters wrote their stories, they used these roles to describe the event to the public. On the basis of the wide recognition and acceptance of these roles, we would say they receive generally wide cultural endorsement.

But what about the roles of "celly," "short," "cop," and "snitch?" These are roles recognized and used by women in the prison. For the most part, however, they receive only limited cultural endorsement. Although they are accepted as real and authentic roles within the subculture of the prison, and are essential for accomplishing certain practical ends on the inside, their use is generally limited to the prison setting. Cultural endorsement for these roles is not widespread. Cultural endorsement, then, can be thought of as a continuum that reflects the degree to which schemas and resources are actually recognized and used as a role. With wide cultural endorsement, the role is seen as almost natural and as quite real. With limited cultural endorsement, the validity of the role is questioned. As I will illustrate later in the chapter, the degree of cultural endorsement associated with a role conditions the frequency and nature of its use as a resource.

Cultural Evaluation

Roles also vary in terms of the degree to which they are seen as either positive or negative. At one end of this continuum are those roles that are highly valued and are associated with prestige, respect, and dignity. At the other end are those that are devalued and are viewed with contempt and disdain. Cultural evaluation is largely independent of cultural endorsement. Roles that are widely accepted as legitimate and real can be seen as either positive or negative. In the case of the prison uprising, this is clearly evident in the role of "inmate," which in most people's minds is linked to the more particular categories of "murderer," "prostitute," "thief," and so forth. These are widely recognized as legitimate, culturally endorsed roles with clear negative evaluations. The role of prison guard, on the other hand, is also widely endorsed, but is generally associated with a more positive evaluation.

Cultural evaluation does of course vary in meaning across groups. The role of prison guard is seen in very negative terms by prisoners while the role of inmate is seen as more positive by those who are incarcerated. In circumstances where roles are evaluated more positively, they are more effective resources for gaining power, dominating others, and directing change. Clearly, then, the context within which roles are being used will condition their utility as a resource. In this sense the women protesters at the prison faced a dilemma in their attempt to produce change. The role of protester could be expected to be viewed in positive terms by other inmates, but in more negative terms by prison staff. If the issue were simply staff vs. prisoners, the staff would win. The women needed an ally and they recognized the importance of the public in their quest (this is evident in the strategy to gain coverage by the press). Yet, the public generally holds negative evaluations of inmates. To be effective at gaining public support this dilemma would have to be overcome.

Social Accessibility

To say that a role is defined by schemas, is to acknowledge that the role is accessible as a resource or available for use, in at least a limited way, by all members of the community that recognize the role as legitimate. This means that a role with high cultural endorsement, or one that is widely accepted as real, is potentially available for use by all members of the community. However, while the community might share equally in their acceptance of the legitimacy of the role, the particular use of the role is often severely restricted. That is to say, social access to the role is variable. For instance, in the declaration of protest written by the strike participants, reference is made to the "Governor of the State of Oregon" and the "Director of Corrections." This constitutes role use in a very broad sense. In this example, these two roles are being used by the inmates to organize their demands, to think about solutions, and to negotiate for change. In this limited sense, we can see how even the role of governor is accessible to all community members. But this does not mean, that all members of the community can use the role of governor for the same purposes. Indeed, only one man could use the role for self-definition in 1988—Neil Goldschmidt.

Again, the point to be made here is that even though a role receives widespread cultural endorsement and is accessible for use by all at some level, all roles are not equally accessible for all tasks. There are a variety of potential uses for a role (a point to be developed in more detail shortly), and some uses are more highly restricted than others.

A somewhat unique quality of the role of protester is that it is potentially available for use to define the self by almost any actor. However, because there are often severe consequences that result from enactment, its use to define the self is generally rare. Only 28 women chose to enact this role in the prison uprising even though it was accessible for this particular use to over 100 women.

Situational Contingency

Roles also vary in terms of when and how they can be used. To the extent that roles require a particular context for use, we can say that they are roles with high situational contingency or are situation *dependent*. Some roles, however, are less likely to require particular settings and in this sense are situation *independent*. Most of the roles employed during the prison protest were situation dependent. For example, one could not enact the role of inmate while outside the prison walls. Similarly, members of the TERT squad would find it very difficult to enact their roles in a setting other than that of an Oregon prison. But at the same time, it is important to recognize that there are also situation independent roles being enacted. Because these roles are by definition ubiquitous, they are often taken for granted. Gender, for example, was clearly an important resource during the uprising, but because it is used as a resource inside and outside of the prison, its use is not as obvious.

Situational contingency has particularly important implications for understanding variation in social accessibility. Roles that are more situation independent will tend to be more accessible, while roles that are dependent upon settings will have more limited access. This is because situations and settings are resources in themselves and can be employed as a strategy for limiting access to other resources. Clearly, the setting of a prison, while ostensibly constructed to restrict certain behaviors and punish convicted offenders, also has the larger effect of limiting access to many other roles. The roles of mother, spouse, and student for example, are difficult to enact in this setting. Even the role of protester, while certainly available, is challenged by the setting since the public may find such a claim to be particularly illegitimate on the inside. Ironically, the role of "rioter" is a more legitimate role inside the prison. This is what the TERT squad prepared for. This is what the public expects and this is what the protesters recognized as available for use.

Variation in Role Use

Variation in role *type* reflects variation in schemas that define social structure. Variation in role *use*, on the other hand, reflects variation in the resources that define social structure. Just as schemas and resources are constituent of each other, so too are the dimensions of role use and role type. Therefore, it is important to keep in mind the contingent nature of role use. Because the situations and combinations that define the social world are highly varied, the potential uses of a role will also be diverse. Nevertheless, there are some general categories of use that have important sociological relevance.

The Definition of Self and Other

A fundamental principle of symbolic interactionism is that roles are employed as categories for defining self and other. These recognizable role identities provide actors with a sense of security, purpose, and direction and also serve to provide a sense of meaningful continuity across situations. To the extent that roles are used to define self and other in mutually acceptable ways, social action is more stable and predictable. In the day-to-day operation of the women's prison, actors usually make claims of self and of other that are not challenged. The prison staff use the role of "inmate" or "convict" to define the incarcerated women and the women use the same role to define themselves. This is not to say that this is the only role used to define the women prisoners, nor does it assume that the women prefer to define themselves as inmates. This co-occurrence of definitions reflects in an extreme form the limits of agency and the effect of a static and powerful structure.

When the definitions of self and other do not coincide, negotiation over competing definitions will emerge. Such a debate is a challenge to the stability of social structure. When these differences become organized among groups of actors who share equally in the battle over self and other definition, social movements to facilitate social change may emerge (cf. Howard 1991). Women, for example, in their historical struggle for equality, have at an important conceptual level advocated for cultural endorsement of the use of certain roles for defining self. Woman as "voter" or woman as "priest," are examples of debates over the use of roles for self-definition. We can see this in the prison uprising where inmates participating in the sit-down strike defined themselves as "peaceful protesters." From the perspective of the prison staff, the women were "involved in an illegal activity," or committing a crime, action that was perceived as consistent with the role of "convict." Thus, interviews with the women who participated in the sit-down strike, suggest that they viewed their action as unrelated to any previous behavior, but the staff, on the other hand, seemed to define the sit-down as consistent with the role of "convict."

Cognition

Implicit in the recognition that roles are used to define self and other, is the more general proposition that roles are also used to think. The idea here is that roles serve as a particular type of cognitive structure. This is a view consistent with the work of social psychologists who study cognition in terms of schemas (Markus 1977), prototypes (Cantor and Mischel 1979), and social representations (Moscovici 1981). Role-thinking takes us beyond these explanations, however, in that it uses social categories that are linked to social structure (Callero 1986).

When roles are used as a cognitive resource they affect how the world is perceived and mentally structured by the actor. This can be seen, for example, in how the accounting of the events of the protest differ significantly among witnesses. One reporter who covered the story recalled that he thought the women were "just looking for an excuse to stay outside on a nice day," while other reporters clearly conceived of the event as a response to prison conditions.

Action

Roles are used to act. This is a proposition that follows from the recognition that roles are used to think and to define self and other. It is also an idea that is quite consistent with traditional structural interpretations of role as a set of normative expectations. But role defined as an element of social structure and as both schema and resource, does not assume action to be the deterministic consequence of role expectations. Rather, role enactment involves relative degrees of discretion and agency. Clearly, the sit-down strike cannot be explained as a consequence of meeting the expectations of the inmate role. On the contrary, this collective act of disobedience was in deliberate opposition to the formal role requirements. The women used a variety of roles to construct the strike. For example, their action employed the roles of protester, citizen, woman, and friend, in organizing and sustaining their brief challenge to authority. Of course, the use of these roles to enact the protest was not without difficulty. Role enactment is always embedded within a context of role use by others. In this sense, individual agency can be limited or structured by the actions of others.

Politics

Roles are a political resource in the sense that they can serve as a tool for exercising control and power. Securing power and controlling others usually involves all of the dimensions of role use identified earlier. Ultimately, however, the most important strategy for gaining influence and exercising power is found in the use of roles as resources for either altering or sustaining social structure. In the dynamic process of using schemas and resources, both stability and change are possible. Through the success of traditional role use, dominant schemas and resources are reproduced and the status quo prevails. But when roles are used in unanticipated ways or when roles from one structure are introduced and used in a novel way in a different social structure, social change is possible. In part, this is what occurred during the sit-down strike. The women prisoners and the prison staff were engaged in a battle over structure and social roles were used to both limit and advance social change.

Social Roles and Social Change

When social roles are conceptualized as resources, we can begin to see how roles can serve as both a source of individual agency and as a means for generating structural change. To appreciate the way in which roles can be used to generate social change, however, requires a somewhat novel theory of structure. In this sense I agree with Sewell (1992, p. 16) when he argues that:

a theory of change cannot be built into a theory of structure unless we adopt a far more multiple, contingent, and fractured conception of society—and of structure. What is needed

is a conceptual vocabulary that makes it possible to show how the ordinary operations
of structures can generate transformations.

I also agree with Sewell's characterization of society as being derived from a
combination of many distinct structures. The structures that make up society
exist at multiple levels and reflect practices derived from sets of schemas and
resources that differ in important ways. Thus, the schemas and resources that
define religious structures differ significantly from those that define educational
structures, which differ still from economic structures, and so on. This is not
to say that the multiplicity of structures that define society do not overlap.
It is very clear that schemas and resources that constrain and enable at one
structural level can also be employed at another. Practices learned in the family
can also be found in practices at the school, for example. The point here is
that while schemas and resources combine in unique ways to reproduce a given
structure, they are not necessarily limited to use in only one structure.

If we conceptualize roles as sets of schemas and resources, as I have
suggested, we can get a more specific understanding of one way in which the
larger structures of institutional spheres might overlap. To the extent that a
role has wide cultural endorsement, is situation independent, and is socially
accessible, it has the potential for being used across structural levels by large
numbers of people. Gender roles, for example are used to reproduce sets of
relations in most, if not all, institutional spheres. In this section I want to show
how the importing or generalizing of roles from one structure of common use
to another structure is key to understanding social change.

This idea builds on the theory of structure developed recently by Sewell (1992)
to explain, in part, how social change can be generated by the normal operation
of structures internal to a society. For Sewell, agency is defined as the ability
of an actor to transpose a schema—that is, to apply it creatively, in an unfamiliar
context. The enactment of schemas in unfamiliar settings creates unpredictability and the potential for change in the reproduction of schemas and the
accumulation of resources. Support for this principle has most recently been
demonstrated by Robin Stryker (1994). In an analysis of legal and scientific
rationalities, Stryker shows that the generalizing of scientific schemas into the
domain of law can account for specific types of change in the legitimacy of legal
institutions. Following Sewell and Stryker, then, I want to suggest that attempts
to change the structure of a prison, may be enhanced by collective acts of agency
in which social roles are transposed or imported from outside the prison
structure. The use of novel roles may create unpredictability and provide new
opportunities for action. At the same time, however, new roles should also lead
to the emergence of a new set of constraints, thus, the importing or transposing
of a role does not guarantee structural change.

When using this framework to examine the sit-down strike it is necessary
to ask: "what were the key roles in the uprising and how were they used?" This
is of course an empirical question. It is also a very challenging question in that
roles are conceived here as existing on a continuum of "cultural endorsement"
which essentially means that the identification or recognition of a repertoire
of social roles is a fluid and often debatable process. I have discussed some
of this in general terms already. I will now take a more in-depth look at one
role in particular, that of "protester."

The Protester Role

At one level, the social structure of the prison is clearly evident in the use
of certain formally articulated roles. From an outsider's perspective, roles of
inmate and prison staff obviously dominate interaction. Within the prison,
more particular roles are also used. For prison staff, roles based on formal
rank, job assignment, age, and gender are important. Among the prisoners,
roles based on criminal offense, time completed, time remaining on ones
sentence, age, race, and sexual orientation are important. All participants in
the prison use these roles to reproduce a relatively stable institutional structure.[6]
However, the role of protester—especially peaceful protester—is not part of
this institutional sphere. It is not part of the normal, everyday orienting
vocabulary of prisoners, correctional officers, or administrative staff. There
is no evidence that any of the inmates use this role as a personal identity, in
other words, it is not a part of the stable self structure of the women. I also
want to argue that the protester role is not expected or anticipated by the prison
administration, indeed, there are only limited historical instances where it has
been used in such a setting. This is not to say that prisons do not experience
forms of protest. Prison riots and other disturbances do occur and are expected
and prepared for by the prison staff. But I want to make the subtle but
important distinction between the role of "protester" and the role of "rioter."
Riots are generally violent, spontaneous, and unorganized. Prison staff receive
training in responding to riots. Special response teams composed of select
personnel prepare for riots. They have physical resources for controlling rioters
and both the prisoners and staff, as well as the public, make use of cultural
schemas for understanding the role of rioter. Most communities will recognize
cultural stereotypes of violent inmates who barricade themselves in a wing of
the prison, take hostages, and destroy property. Even though this is not a
common occurrence, it is nevertheless, an aspect of the cultural schemas that
contribute to the structure of the prison. In this sense, the role of rioter can
be conceived as a resource within the existing social structure of the prison.
While the role of "rioter" is certainly not as fundamental as that of "guard,"

"snitch," or "warden," it is certainly more important and central than that of "protester." Because the role of protester is not part of the prison social structure, when it was used in the uprising at the women's prison it had to be transposed from a structure outside of the prison setting. Moreover, because it was an *imported role*, it was a potentially more effective resource than the role of rioter. To the extent that the role of rioter is maintained within the institutional sphere of the prison, prison structure is less likely to be altered by its use. Peaceful protests, on the other hand, are not anticipated, are more likely to be inconsistent with prevailing schemas, and are therefore reflective of a greater degree of agency. Riots, therefore, may actually be less effective in changing structure than peaceful protests.

Variation in Type

We can say that the role of protester has a generally high level of *cultural endorsement*. It is a role that is clearly part of the stock of common knowledge in our culture. While the meaning of this role may not be highly crystallized for most, it is certainly widely recognized and widely used as a resource for understanding certain types of action. On the other hand, it can also be said that the role of protester, while having wide endorsement, also has a relatively high level of *situational dependency*. This is evident in the fact that the definition of self or other as protester really only makes cultural sense in the context of an actual protest. Unlike the roles of, say, female, mother, or Lutheran, one cannot claim the protester role very easily unless the immediate situation supports such a claim.

This does not mean, however, that the protester role has limited *social access*. In fact, it is, in theoretical terms, a highly accessible role. There are very few social prerequisites of enactment and almost any actor can "protest" at any moment. Yet enactment is rare. This is due in part to the realization that the consequences of enacting the protester role are often severe. It is also due to a perception that the protester role has a generally negative level of *cultural evaluation*. For a variety of different reasons, a protester is perceived as deviant. Generating a positive evaluation depends on support for the legitimacy of "the cause." In fact, the wide collective use of the protester role, and the subsequent success of the protest itself, depends to some degree on the protesters' ability to sustain a positive cultural evaluation during role enactment. Single acts of protest are less effective than collective acts and collective acts with wide public support are more effective than those with limited support.

Variation in Use

Identifying the specific levels of endorsement, access, evaluation, and situational dependency are critical because these dimensions serve to structure the particular uses of the role as resource for definition, thought, action, and politics. Although it is possible to identify analytically distinct dimensions of role use, it is important to keep in mind the fact that these uses overlap in various ways. This can be seen, for example, in the use of the protester role as a source of identity. The women who engaged in the sit-down strike used the protester role to help define themselves. But importing or transposing this particular role into the institutional setting of the prison was risky. Would the others in the setting use the protester role to define the defiant women? This is a critical question because the success of the protest and the ability of the women to sustain the protester identity, as well as their own personal safety, depended to a large extent on whether other prisoners, staff, media, and members of the public also defined the women as protesters. While a shared definition did emerge among some actors, a larger consensus in definition did not occur. One prisoner who did not participate in the sit-down, reflected recently on her perception of the protesters:

> They—they needed to be more organized for—to get me to do this. I couldn't understand what they were fighting about. It gave them something to do. A lot of girls, they just did it 'cause it gave them something to do and they fit into a little group.

For this former inmate, the decision not to join was associated with a perception of the women as a disorganized group motivated by "non-protester" reasons.

It is also clear that the protester role was not widely used by the prison staff to define the women. The written reports describing the incident that were submitted by all participating correctional officers, consistently defined the women in terms inconsistent with the role of protester. Of the 25 reports that I was able to gain access to, none of the descriptions used the words "protest" or "protester." Instead, the incident was consistently referred to as the "illegal activity," "disturbance," and "emergency situation." Some prison guards did make reference to the "sit-down," but the participants in the sit-down were never classified using a similar vocabulary. Instead, the institutional role of inmate was the dominant category for classifying the women protesters. The following excerpt from the report of the security manager is representative (emphasis added):

> They informed me there were approximately 20 *inmates* in the yard involved in a *sitdown*. At approximately 4:27 we ordered yard in. Officer D—- ordered this by saying, 'Yard in, Yard in. *Inmates* report to your cells.' Approximately 40 *inmates* at this time came inside the institution. 28 of the *inmates* did not.

This is not to say that the prison staff did not recognize the actions of these inmates as composing a unique collective act. This is clearly the case. However, evidence from official reports and retrospective accounts, suggest that "rioter" was the dominant role used by the prison staff to define the women who took part in the sit-in. This is suggested in the comments of the same security manager who was interviewed by a local reporter two weeks after the incident; she said:

> We do not negotiate under threat. I don't have a crystal ball that can tell me that just because a sit-in starts out peaceful, it will remain that way. There are proper channels for grievances.

Thus even though a "riot" did not occur, the women were defined as "potential rioters" by prison officials.

As members of the prison community, the women who participated in the sit-down were well aware of the discontinuity in the role they were using to define self, and the role others might use to define them. The written declaration of protest reflects this "... if not for this peaceful demonstration the likelihood of a full scale riot would be realized." In a similar statement, one inmate participant noted in a letter to a reporter a day after the incident that:

> Talk has always been, 'We should riot. We need to take over this place, they can't run it right.' Well, they were talking about a riot again....Some sensible people decided a peaceful demonstration with no violence just might wake the administration up.

This view is also consistent with the following statements reportedly made by two different women while they were being pulled from the circle by correctional officers. Recognizing that the peaceful protest was coming to an end, they threatened violence: "Remember we tried it peaceful" and "Let's tear this place apart. What do we have to lose" (quoted from two separate officer reports).

Although it is clear that prison staff did not use the role of protester, the print media that covered the event did use this role in their public news reports to describe the women. Note the following headlines that appeared in local newspapers: "Inmates Protest Crowded Prison," "Women Inmates Protest,"

"Inmates Punished for Protest," "Six More Inmates Will Pay for Protest at Women's Prison." Within the news articles the women were never referred to as "rioters" and the incident was never described as a riot. Instead, "sit-down," "disturbance," "sit- in," "demonstration," and "protest" were consistently employed.

The fact that the media used the protester role was a victory of sorts for the women. The public's understanding of the incident would be determined in large part by the media's interpretation and explanation of the events. If the public defined the women as protesters, the women's claims would more likely be given a greater degree of legitimacy.

When roles are used as a resource for defining the self and other, they are also used as a resource for action and thought. This was clearly evident in the strategies and actions of the women protesters. Before enacting the role of protester, organizing and planning occurred among certain leaders and recruiters. The actual "thinking" of the women is obviously impossible to observe, but the articulation of the grievances and demands in the written note announcing the demonstration gives us some insight into how the event was conceived prior to enactment. It is clear, for example, that the women hand in mind a "quiet, peaceful sitdown demonstration" and a "Hunger Strike." They also anticipated an official response that could produce "violence and or...serious Disciplinary Actions." This suggests an understanding of how the protester role would be enacted and how it would be perceived by prison officials.

Enactment was structured by all participants understanding of the role of protester, which is part of the cultural stock of knowledge as well as more specific personal experiences. A prison guard noted in her report that she had overheard one inmate joking to another by saying "she wished she had a record of Joan Biaz [sic]." This statement suggests an understanding of the protester role in terms associated with the civil rights and antiwar protests characteristic of the 1960s and 1970s. Although the setting was a correctional institution in 1988, the enactment of the protester role was quite consistent with popular images of protest from this earlier era. This includes the occupation of a public space (the prison yard), the linking of arms while sitting, the singing of protest songs, the use of fasting, and the specification of one's political position in the form of a written set of demands. All of these acts are recognizable outside the prison as the behavior typical of a protester.

Many of the women who did not participate in the sit-down served as an audience that could validate the protester role. For example, while being escorted to their cells, a number of women raised a clenched fist in a symbol of solidarity and defiance. This was met with cheers of support from the other prisoners. Similarly, once the protesters were segregated, they were collectively referred to by the larger population of prisoners as "the 28 they hate," a phrase that may have been modeled after similar characterizations of more famous groups of political prisoners (e.g., "The Chicago Seven" and "The Gang of Four").

In sum, there is evidence suggesting that the protester role was used as a resource for self and other definition and as a tool for organizing both thought and action. But perhaps the most critical use of the role may have been as a resource for achieving certain political ends. The women protesters could be viewed as struggling to change the social structure of their prison. I have argued that they implicitly recognized that they could *not* do this by using the role of inmate. In other words, it was obvious to some women that change was not going to occur unless they took the initiative and acted in ways not typical of rule-following inmates. Thus, they had to do something different, something unexpected. In theoretical terms, they had to draw on resources from outside the prison. Importing the role of protester served this end. The women obviously did not intend to incorporate the protester role into the more permanent social structure of the prison, this would be culturally inconsistent with the meaning of protester. Nor were they seeking to establish a radically new set of relations, this is clear from their list of grievances. Instead, they wanted relations in the prison to change in very small but personally significant ways. To accomplish this they tried to use the protester role to gain access to other resources for producing change.

The media was the most important of these. The women appeared to understand that the protester role could give access to the media and the essential public forum. Thus, newspapers were called *before* the demonstration was constructed. Based on a similar logic, the protester role was also used to try and establish access to the Governor of the state of Oregon, the Director of the Department of Corrections, and the Superintendent of the prison. The women believed that all three of these roles could potentially be used to produce the changes sought by the protesters. The women were thus empowered by the protester role. They used it to enhance their ability to act as agents within a structure that places extreme limits on agency. In this sense, the protester role was a resource for social change.

The irony and dilemma of roles, however, is that while they are essential tools for challenging structures and enabling creative action, they also simultaneously serve to structure and limit action. As "peaceful protesters" the women were committed to a line of action made possible by the protester role. The protester role helped organize and coordinate a collective act of defiance. In using this resource, however, other possible acts were effectively cut off. Acts inconsistent with the protester role could not be used without altering others' definition of the women as protesters.

In a similar way, the protester role structured action through the schemas it represented (what I have called variation in role type). Wide cultural endorsement enhanced recognition by the public and facilitated the organization of the uprising. A high degree of social access made it a reasonable resource for constructing change, but the tenuous level of cultural evaluation meant that public actions and protest demands had to be framed in a manner acceptable to the public. The women, for example, could not use the role to demand freedom from prison—this they understood. Finally, the situational contingency of the protester role meant that once the women were removed from the public space in the yard and isolated in separate cells, use of the protester role would be difficult if not impossible.

CONCLUSION

This role analysis is necessarily incomplete. It is incomplete in that I have not explored the larger structural and historical context within which the uprising occurred. It is also incomplete in that I did not analyze the use of key roles beyond that of protester. Finally, I have not discussed the consequences of the protest for the women and for the structure of the prison. Unfortunately, space does not permit an extensive exploration of these issues. I will, however, briefly comment on each of these points by way of a conclusion.

First, the protest occurred at a time of relatively severe overcrowding in the Oregon prison system. The women's prison, for example, was at almost 200 percent of capacity and, according to the U.S. Bureau of Justice, was the second most crowded women's prison in the country. This meant that recreation space in the building was being converted to rooms for additional beds, which was clearly perceived by the inmates as a removal of longstanding privileges.

Security had also been increased dramatically over the year. The movement of prisoners became more highly controlled. Cell mates were determined by the staff, gift boxes from the outside were banned, new guards were hired, and the physical structure of the perimeter was reinforced with razor wire. These changes were partly a response to the demands of controlling a growing population and partly the consequence of the highly publicized escape of a famous inmate.

In addition, in the year prior to the protest, changes in the management and organization of the Corrections Department were leading to changes in the day-to-day operation of the prison. There was an administrative push to develop a uniform set of procedures and practices that could be applied across all of the state prisons. This covered everything from educational and recreational offerings to food and prison clothing. The emphasis on uniformity came from two likely sources. One was the bureaucratic ethic that standardization is associated with increased efficiency—a sentiment spurred by the growth in size. The second motive was to avoid lawsuits. A number of men and women had initiated legal challenges to differences in treatment across prisons. The differences, were, and still are, quite evident, especially between the men's and women's institutions. For example, women are required to do their own laundry, individually, while men have a laundry service built into a work program, for which they are paid. Also, men have a more diverse set of job training opportunities in the traditional male fields of woodworking and machine operation. Women, on the other hand, could opt for training as a beautician or secretary, jobs that are more traditionally occupied by women and are lower paying.

The gender difference evident in the treatment of prisoners, also serves to illustrate the relevance of roles other than that of protester. The role of "woman" was clearly an important resource in the uprising. Although I could find no evidence that the protesters thought of themselves as feminists, they nevertheless used the role of woman to organize their thoughts, actions, and identity. In the words of one protester, "I'm not a feminist, I just want to be equal with men." Separating the use of diverse roles for analytical purpose can be quite difficult and at times is impossible. For example, the participants in the uprising were not perceived simply as protesters by other (or by self). They were women-inmate-protesters from the public's perspective and women-inmate-rioters, from the prison administration's point of view. The overlap in use of these roles alters their meaning and structures their use as a resource in significant ways. The argument could be made, for example, that woman prisoners can make more effective political use of the protester role than can male prisoners. This is because the public image of women as generally more nonviolent than men, enhances the likelihood that the public will define women as protesters and men as rioters. And to the extent that men and women use gender roles in a similar way for self-definition, one could argue that women would be more likely to use the protester role over the rioter role as a resource for political action. In fact, it is interesting to note that exactly one month after the uprising at the women's prison, there was an uprising at one of the neighboring men's institutions. The men, however, barricaded themselves in a dormitory room, broke furniture, and were subdued with fire hoses in an incident that appeared to be more spontaneous and disorganized than the women's sit-down.

Earlier I argued that the role of protester was a more powerful resource for changing structure than was the role of rioter. Because the prison system in Oregon was experiencing a relatively high degree of change at the time of the women's uprising, it is difficult to determine what effect, if any, the uprising had on the social structure of the prison. Certain gains in educational equality have been made for example, but these may have been due more to lawsuits than anything else. Also, the head of the Department of Corrections, Michael Franke, was stabbed to death outside of his office four months after the uprising. The murder brought an unprecedented amount of public attention to the prison system as a whole and lead to an official investigation into charges of high level corruption and conspiracy in the Corrections Department. A former inmate at the men's prison was eventually convicted of the offense, but the political ramifications of the murder most certainly worked in an indirect way to alter the structure of the women's prison. Partialing out the effect of the women's uprising in this context is obviously not possible.

It is possible, however, to identify the immediate consequences the protest had on the lives of the 28 women participants. The punishment for participating was severe. Most spent a full year in "segregation," and a few who were perceived as leaders, spent a year and a half. Segregation meant being isolated in their cells with no privileges. Meals were delivered and they were permitted to exercise in a small room for 20 minutes a day. There was no television, no visitors, no library, no education or drug programs, and only limited daylight. The psychological consequences of the isolation were so dramatic for some of the women that a couple were later transferred to the State Hospital Psychiatric Ward.

The length of "seg-time" received by the protesters was perceived as highly excessive by inmates familiar with the sentence standards. Women who engaged in violent disturbances such as fighting and rioting did not receive anything near the seg-time of the protesters. So why were the penalties so harsh? Although there is little evidence to support my claim, I believe that a collective protest is perceived as a greater threat to the prison structure than single acts and that a peaceful protest is even more threatening than a violent disturbance. Understanding why this would be the case is enhanced by an appreciation of roles as resources.

NOTES

1. The following description and subsequent analyses are based on interviews with prison inmates, guards, and newspaper reporters who participated in the event. The analysis also makes use of official institutional reports of the incident. Some names have been changed to help protect the anonymity of the subjects.

2. The text is completely reproduced in its original form. Grammar, punctuation, and spelling have not been altered.

3. The last part of the note was apparently written on the back of the last page. It was a very small continuation of the last point and was not noticed when my source made a copy of this document.

4. Although I see this role analysis as belonging to the larger "resource perspective," I want to emphasize that my particular conceptual approach takes roles as both schemas and resources.

5. The limitations of Stryker's framework in this sense are shared by the similar approaches of McCall and Simmons (1978), and to a lesser extent, Turner (1978). For an elaboration of these issues and the related argument that a resource approach is consistent with the pragmatism of George Herbert Mead, see Callero (1986).

6. By focusing on the protester role I do not want to suggest that these other roles are not critial for understanding the internal dynamics of the sit-down strike. Indeed, it could be argued that gender is the most significant role in this incident. Nevertheless, a complete account must at some level address the role of protester as a resource *along with* the other resources of gender, age, sexual orientation, and so forth.

REFERENCES

Baker, W. E. and R. R. Faulkner. 1991. "Role as Resource in the Hollywood Film Industry." *American Journal of Sociology* 97: 279-309.

Bales, R. F. 1958. "Task Roles and Social Roles in Problem-Solving Groups." In *Role Theory: Concepts and Research*, edited by B.J. Biddle and E. J. Thomas. New York: Wiley.

Bourdieu, P. 1977. *Outline of a Theory of Practice*. Cambridge: Cambridge University Press.

Callero, P. L. 1986. "Toward a Meadian Conceptualization of Role." *Sociological Quarterly* 27: 343-358.

————. 1991. "Toward a Sociology of Cognition." In *The Self-Society Dynamic: Cognition, Emotion and Action*, edited by J. A. Howard and P. L. Callero. Cambridge: Cambridge University Press.

————. 1994. "From Role-Playing to Role-Using: Understanding Role as Resource." *Social Psychology Quarterly* 57: 228-243.

Cantor, N. and W. Mischel. 1979. "Prototypes in Person Perception." In *Advances in Experimental Social Psychology* (Volume 12), edited by L. Berkowitz. New York: Academic Press.

Cottrell, L. S. 1942. "The Adjustment of the Individual to his Age and Sex Roles." *American Sociological Review* 7: 617-620.

Davis, K. 1949. *Human Society*. New York: Macmillan.

Gerhardt, U. 1980. "Toward a Critical Analysis of Role," *Social Problems* 27: 556-69.

Giddens, A. 1984. *The Constitution of Society*. Berkely, CA: University of California Press.

Goode, W. J. 1960. "A Theory of Role Strain," *American Sociological Review* 25: 483-496.

Gross, N. A., A. W. McEachern, and W. S. Mason. 1957. Pp 447-459 in *Readings in Social Psychology*, edited by E. E. Maccoby, T. M. Newcomb, and E. L. Hartley. New York: Holt.

Hewitt, J. P. 1989. *Dilemmas of the American Self*. Philadelphia: Temple University Press.

Hilbert, R. 1981. "Toward an Improved Understanding of 'Role'." *Theory and Society* 10: 207-226.

Howard, J. A. 1991. "From Changing Selves Toward Changing Society." Pp. 209-238 in *The Self-Society Dynamic: Cognition, Emotion and Action*, edited by J. A. Howard and P. L. Callero. New York: Cambridge University Press.

Linton, R. 1936. *The Study of Man*. New York: D. Appleton-Century-Crofts.

Markus, H. 1977. "Self-Schemata and Processing Information About the Self." *Journal of Personality and Social Psychology* 35: 63-78.

Merton, R. K. 1957. *On Theoretical Sociology* New York: The Free Press.

Moscovici, S. 1981. "On Social Representations." Pp. 181-209 in *Social Cognition: Perspectives on Everyday Understanding*, edited by J.P. Forgas. London: Academic Press.

Parsons, T. and E. A. Shils. 1951. *Toward a General Theory of Action.* Cambridge, MA: Harvard University Press.

Schwalbe, M. L. 1987. "Mead among the Cognitivists: Roles as Performance Imagery." *Journal for the Theory of Social Behavior* 17: 113-133.

Sewell, W. H., Jr. 1992. "A Theory of Structrure: Duality, Agency, and Transformation." *American Journal of Sociology* 98:1-29.

Stryker, R. 1994. "Rules, Resources, and Legitimacy Processes: Some Implications for Social Conflict, Order, and Change." *American Journal of Sociology* 99: 847-910.

Stryker, S. 1980. *Symbolic Interactionism: A Social Structural Version.* Palo Alto, CA: Benjamin/ Cummings.

Stryker, S. and A. Statham. 1985. "Symbolic Interaction and Role Theory." Pp. 311-378 in *The Handbook of Social Psychology* (3rd edition), edited by G. Lindzey and E. Aronson. New York: Random House.

Turner, R. 1962. "Role-Taking: Process vs. Conformity," Pp.22-40 in *Human Behavior and Social Processes*, edited by A. M. Rose. Boston: Houghton Mifflin.

―――― . 1978. "Role and the Person," *American Journal of Sociology* 84: 1-23.

White, H. C., S. Boorman and R. L. Breiger. 1976. "Social Structure from Multiple Networks. Part 1, Blockmodels of Roles and Positions." *American Journal of Sociology* 81:730-80.

SOCIAL IDENTIFICATION AND SOLIDARITY:
A REFORMULATION

Barry Markovsky and Mark Chaffee

ABSTRACT

Sociology has focused little systematic effort on understanding one of the most fundamental of social phenomena: the forces that account for and maintain "groupness," that is, solidarity. Our review of three literatures—cohesion, solidarity, and social identity—makes evident some of the problems that have plagued this inquiry. In an initial attempt to introduce greater clarity and rigor to this historically muddied field, Markovsky and Lawler (1994) introduced a structural definition of solidarity that addressed both the relations between individual social actors and the higher-level patterns of those relations. Building from that formulation, the present work opens new conceptual and empirical domains by allowing *social identification* to serve as a relational basis of group solidarity. We develop a number of implications of this reformulation and draw out some of the research phenomena that could be potentially addressed, especially as related to those forces that alter the solidarity of a set of actors.

Advances in Group Processes, Volume 12, pages 249-270.
ISBN: 1-55938-872-2

Sociological explanations of group solidarity and related phenomena are relatively few in number and loosely formulated. Although we are certainly not the first to theorize about solidarity, we find it surprising that in a field so concerned with human group phenomena, there are relatively few others engaged with this problem and even fewer rigorous theoretical developments. Relatively little effort has been devoted to developing and testing abstract and general theories of group solidarity. Our approach is primarily *structural* in that solidarity is defined in terms of relatively stable patterns of social ties. In turn, the content of those social ties is not fixed by the theory. Any binding ties, from interpersonal sentiment bonds to dependency relations to an individual's identification with and attraction to a group, may provide the basis for group solidarity—as long as certain structural imperatives are satisfied.

This approach builds upon the work of Markovsky and Lawler (1994) who departed from traditional definitions by reconceptualizing solidarity in terms of network structure. They treated solidarity as a network property characterized by the strength, duration, and integrity of attractive bonds between members. The approach directed attention to the *patterns* of relations among members of a set of actors. They also developed theoretical propositions about the effects of different *interpersonal* processes on *group* solidarity, and the effects of solidarity on various individual and group outcomes.

After reviewing prior theoretical work, we will introduce several developments that attempt to broaden and deepen our understanding of group solidarity. They do so by establishing or strengthening bridges to various literatures and by further refining the definition of the solidarity concept. Linking the earlier work of Markovsky and Lawler (1994) to social identification theory, we will examine the social structural implications of the psychological phenomenon of group identity. In addition, we will initiate an effort to understand the effects of different structural forms on a group's capacity to maintain its integrity against disintegrative forces. In the course of these discussions we will also take a brief look at how solidarity is affected by network multiplexity—the presence of multiple types of social ties.

PRIOR SOLIDARITY THEORIZING: A CACOPHONY IN THREE LITERATURES

At least three distinct literatures bear on the central topic of this chapter: cohesion, solidarity, and social identification. Sociological work has emphasized solidarity, usually in larger group contexts (Durkheim 1956; Gamson, Fireman, and Rytina 1982; Hechter 1987). Psychological approaches have tended to focus on the empirical analysis of cohesion, usually in small group contexts (Festinger, Schachter, and Back 1950; Lott and Lott 1965; Cartwright and Zander 1968). Social identity theorists such as Tajfel (1972), Turner (1975), and Hogg and Abrams (1988) have attempted to link the theory and empirical findings of cohesion and solidarity. Each of these three strands will be examined in more detail subsequently.

Cohesion

Cohesion has one of the most extensive intellectual histories in social psychology, but it has been plagued by problems of clear definition and limited operationalization.[1] Expanding upon Lewin's (1952) work in field theory, Festinger provided what has since become the most frequently cited definition of group cohesion: "the resultant of all the forces acting on the members to remain in the group" (1952, p. 274). He and others believed that cohesiveness is determined by the attraction of members to the group and the extent to which the group mediates important goals for its members.

Field-theoretic roots notwithstanding, theorists have usually treated cohesion as an intragroup phenomenon, operationalized in such a way as to reduce Festinger's (and Lewin's) "all forces" to one: inter-member attraction or liking (see Cartwright and Zander 1968; Hogg 1985, 1987; Lott and Lott 1965; McGrath and Kravitz 1982; Turner 1984; Zander 1979). Lewin's (1948, 1952) field theory suggests that group cohesiveness is a quality that varies "from a loose 'mass' to a compact unit" (1948, p. 84), determined by the psychological representation in the individual's "life space" of inter-individual forces of attraction and repulsion within the group. Though Lewin's life space conceptualization invites many different operationalizations of the concept of cohesion, it has instead typically been treated, both theoretically and operationally, as interpersonal attraction. Cohesiveness of the group is then measured as the average magnitude of all dyadic, inter-member attractions. Moreover, conceptual and measurement problems associated with holistic assumptions (e.g., the sum of *all* forces...) are not only daunting, but quite probably unsolvable (e.g., Nagel 1979, pp. 380-397).

Concomitantly, cohesion theorists have treated inter-member attraction as an absolute prerequisite for the existence of a group (Bonner 1959, p. 66; Shaw 1981). They believe that this attraction fosters interactions that satisfy needs, drives, motives, or desires. It then follows that people form groups when doing so provides them with opportunities for mutual needs-satisfaction. In this way the establishment of the group promotes mutual dependence. The interdependence of group members for the attainment of goals covaries with group cohesiveness. The more interdependent the group's members, the more needs get satisfied, and the more cohesion in the group.

Festinger, Schachter, and Back (1950), Festinger, Reicken, and Schachter (1956), and Heider (1958) also emphasized that similarity is frequently a basis for interpersonal attraction and thus a basis for group cohesion. Festinger's social comparison theory proposed that people seek similar others to help gauge their own perceptions, beliefs, and social knowledge. More than dissimilar others, those who are similar are more likely to provide validation for these inherently subjective elements. In Heider's well-known theory of cognitive balance, one basis for stability among a set of individuals is perceived similarity across members' attitudes and attributions regarding one or more situational elements such as beliefs, persons, or groups. Given balanced relations, forces for group structural change are minimized resulting in stable, cohesive sets of actors.

To what extent is interpersonal attraction synonymous with or related to the liking of the individual for the group? Though often conflated in the cohesion literature, there is no a priori reason to presume synonymy. It is not difficult to conceive of situations where members of a group like being members and are attracted to the group in spite of a dearth of close personal ties to other members. Conversely, members may even dislike their group as an objectified unit while still enjoying positive relations with many other members. Further, when we consider sources of attraction other than positive sentiments toward specific others, for example, the prestige of the group or the nature of its activities, research has generally indicated that these factors are not significantly related to traditional indicators of cohesion such as group performance and task effectiveness (Eisman 1957; Jackson 1959; Ramuz-Nienhuis and Van Bergen 1960). Such findings show that different operationalizations of cohesion lead to conflicting findings, thus raising questions concerning the adequacy of interpersonal attraction as a proxy for "all forces."

Variations and ambiguities in working definitions for the concept of "group" have further muddied cohesion's theoretical waters. For example, Bonner (1959) treated attraction among members as necessary and sufficient for group formation. If attraction is the causal process leading to psychological group formation, however, then what is the group if it is not, tautologously, the set of mutually-attracted actors? Lacking a clear definition of the group that is independent of its determinants, cohesion theorists are, at the very least, still faced with either explaining the origin of groups or more clearly defining them. Additionally, cohesion *qua* interpersonal attraction is frequently assumed to be based on shared goals, beliefs, values, or attitudes. This presupposes that people come into each other's environments already possessing such things. Yet we know that groups play dominant roles in shaping goals, beliefs, values, and attitudes. The theorist's task then seems to become specifying conditions under which (1) coincidences across actors' goals, beliefs, and so forth, are sufficient to trigger the collective investments in time and energy that are needed to create particular groups, and (2) a particular set of goals, beliefs, and so forth becomes more homogeneous in groups once formed. It will not do to make the universal claim that group membership is based upon what is shared, or that what is shared becomes so due to group membership.

A related problem for cohesion theories is the treatment of group formation as owing to attraction between members of nascent groups, in turn based on mutual dependencies for goal attainment. If members are attracted to each other as goal mediators, then we should expect that as groups fail, members' attraction to the group should diminish and, ultimately, the group should dissolve. This prediction does not hold in many situations, however. Festinger et al.'s *When Prophecy Fails* (1956) is an account of how groups, in failure, commonly find new intragroup resources that permit them to carry on. Rather than becoming weakened by their failure, they are instead galvanized and their cohesiveness heightens. Experimental evidence supports these observations. Turner, Hogg, Turner, and Smith (1984) found that groups that failed at an experimental task expressed greater cohesiveness along a number of different dimensions compared to successful groups. Similarly, Turner, Sachdev, and Hogg (1983) found that unpopular groups—groups that were consensually disliked in the experimental context—were more cohesive than popular groups.

One of the fundamental questions posed in the cohesion research concerns the limiting effect of group size on cohesion. With interpersonal liking as the primary basis of cohesion, there must be a size limit to groups that may be deemed cohesive because each member can only have a relatively small number of strong interpersonal ties with others. Every strong tie requires interaction for its maintenance, and we simply haven't the time to engage in more than several such relationships. For any individual, such interactions must be limited in their number, quality, and intensity. With increasing group size, maintaining relations from each to all others becomes problematic. Increasing group size then encourages the formation of subgroups. As subgroups increasingly take over functions that were formerly provided by the larger group for its members, they diminish the cohesiveness of the larger group (Gerard and Hoyt 1974; Porter and Lawler 1968; Markovsky and Lawler 1994).

The issue of group size further complicates theory development by generating seemingly contradictory effects. Research findings have shown that while increasing group size may limit the proportion of members with whom one may have strong relations, it can also magnify the impact of the group on the individual, strengthening compliance to norms and producing more group-directed behavior (Latané 1981; Paulus 1983). If we treat conformity and cohesiveness as indicators of "group belongingness," then we are faced with the need for a theory that can encompass the difference between the ties that bind the small group and those that bind the large group.

Solidarity

More than either psychologists or social psychologists, sociologists have focused on the impact of structural dynamics on groups and on the sense of "belongingness" at the large group level. For reasons not entirely clear, they more readily label their interest as "group solidarity" rather than group cohesion. The inquiry into group solidarity was established most explicitly by Durkheim (1956/1893). He posited two forms: mechanical and organic. Mechanical solidarity occurs in groups of people who share primary origin characteristics such as language, geography, family, or community. He contended that such characteristics tie people together through strong emotional bonds and interwoven social norms. Such bonds and norms are expressed and experienced as a "collective consciousness." This form of solidarity is common among smaller, preindustrial societies.

Organic solidarity, in contrast, exists among groups of people who are tied together by functional interdependence due to increasing divisions of labor and the specialization of labor that results. In the modern industrial and postindustrial world, this interdependence has replaced the origin-similarities of mechanical solidarity, thereby reducing the collective consciousness and its concomitant role in maintaining group memberships. Durkheim viewed solidarity as diminishing with the rise of science and technology in a society.

A more recent perspective on solidarity based on rational interdependence was proposed by Hechter (1987). He offered that solidarity is a function of (1) corporate obligations and (2) the likelihood that members will comply with those obligations. He also referred to these two dimensions as "defining properties" of solidarity (1987, p. 18), while later asserting that solidarity varies with (i.e., is determined by) members' investments in the group (1987, p. 168). Similar to cohesion theorists, neither Hechter nor Durkheim offered a definition of solidarity that was explicitly distinguished from its determinants and consequences.

Hechter saw his analysis as moving the discussion of solidarity out of the nebulous realm of normative analysis—a nebulousness that he felt could not adequately address the large number of "free-riders" in groups with normatively-based solidarity.[2] It remains to be seen, however, whether his notions of "obligation" and "likelihood of compliance" are any less obscure and more theoretically powerful. Do sanctions for failing to meet obligations differ from sanctions for violating norms? We believe that with attention to clearer definition of some key concepts, Hechter's argument may become quite compelling. However, we are also doubtful that obligation-based compliance under threats of group sanction is the exclusive path to solidarity.

Hechter's formulation suggests that positive, supportive behavior on the part of group members is essentially coerced by other members via threats of shame and punishment. Even if this is so for some who contribute to the group, it is certainly not the only basis for group-oriented behavior. Further, the Hobbesian egoist that underlies Hechter's model of solidarity portrays the group as seeking to extract from the individual something that she or he would not otherwise provide. Lost is any possibility of voluntaristic sacrifice and joyful membership.

Markovsky and Lawler (1994) proposed an explicit conceptual linking of cohesion and solidarity. Central to both was the concept of *reachability*: the relative closeness of network members in terms of both direct *and indirect* attractive bonds. The authors then defined cohesion in terms of reachability, that is, a set of actors was said to be cohesive to the degree that it has high reachability. The definition for *solidarity* further required the concept of *unity of structure*—the relative absence of cliques or subnetworks. By defining solidarity in such abstract and structural terms, Markovsky and Lawler managed to disentangle the concept of solidarity from potential social and psychological determinants and consequences. The formulation was an effort to link interpersonal psychology and group structure, offering a glimpse into a possible unification of small group and large group research as well as a unification of utility-based and nonutility approaches. Unlike much of the prior cohesion and solidarity theorizing, their work also offered clear theoretic assumptions, making possible the derivation of explicit, testable hypotheses. Later we will return to their formulation.

Social Identification

Social identity theory provides an alternative entry into solidarity and cohesion phenomena (Tajfel 1972; Turner 1982). One of the central problems of this approach is to account for the formation of groups. Group formation, in turn, has as necessary and sufficient conditions the self-categorization or self-identification of an individual into a category shared by one or more other individuals. Further, by the definition of "group" in this theory, all relevant actors are aware of the collectivity they so form (Turner, Hogg, Oakes, Reicher, and Wetherell 1987). Group membership is based on identifying characteristics of the individuals involved, characteristics that could include similarity of attitudes, physical or cultural characteristics, or experiences.

Social identity theorists have criticized social cohesion explanations, arguing that they deal with an interpersonal phenomenon—liking—rather than with groups, as claimed by cohesion researchers. They believe that cohesion approaches ask the wrong questions. Rather than studying how groups come into existence and what internal and external factors affect their strengths, cohesion approaches consider what makes people socially attractive to each other and the consequences of such attraction in group settings. Social identity theorists challenge their cohesion counterparts to account for group members' presumed ability to "contain psychologically the whole within themselves; that is, [how] they can cognitively represent the group to themselves and act in terms of that cognitive representation" (Hogg and Abrams 1988, p. 101).

Although also known for making inroads in our understanding of intergroup processes, social identity theory is also notable for its assumptions about intragroup processes. According to Hogg and Abrams (1988, p. 93), intragroup behavior "refers to interaction between two or more individuals that is governed by a common or shared social self-categorization or social identity." Several explicit definitions central to this approach include:

- *Social Categorization:* the discontinuous classification of individuals into two distinct groups, an in-group and an out-group (Hogg and Abrams 1988, p. 51.
- *Self-Categorization:* cognitive groupings of oneself and some class of stimuli as the same (e.g. identical, similar, equivalent, interchangeable) in contrast to some other class of stimuli (Turner et al. 1987, p. 44).
- *Social Identity:* the individual's knowledge that he belongs to certain social groups together with some emotional and value significance to him of the group membership (Tajfel 1972, p. 31).
- *Social Group:* two or more individuals who share a common social, self-categorization and perceive themselves to be members of the same social category (Turner 1982, p. 15).

Social identity theory's approach to solidarity thus deviates from the general thrust of social psychology: "...solidarity and group cohesiveness is not just a matter of shared goals and aims but also something which secures allegiance to the group rather than to individuals in the group" (Hogg and Abrams 1988, p. 92). The theory offers that one's self-categorization with a set of others perceived as similar explains the phenomenon of group belongingness and its consequences. It considers the degree and context of interpersonal similarity in accounting for variations in intragroup behavior.

Within the social identity framework, group membership is a mental state entailing a shared cognitive representation among social actors—a representation that informs each individual of who she or he is and what she or he is to do. In the social presence of others, individuals categorize themselves as members of a group, similar or dissimilar to those in their social presence. Social and self-categorizations do not create the category on the basis of interpersonal attraction, but rather based on shared identifications.

By accounting for group formation via self-categorization, social identity theory distinguishes group belongingness from interpersonal attraction. In so doing, it sidesteps the problem of accounting for all of the "attractive forces" that might pull people together to form a group, as well as the problem of defining the cohesive group in terms of size limitations. Through the categorization process, the theory accounts for two different attractive forces: (1) social attraction, such as, liking between individuals due to shared group memberships (Hogg and Abrams 1988, p. 107), and (2) interpersonal attraction, such as, liking between individuals based upon their unique, individual characteristics and differences. It is the distinction between these two attractive forces that allows us to consider a new dimension in our solidarity theory.

NETWORK SOLIDARITY

The remainder of this chapter is devoted to clarifying, extending, and exploring the consequences of a network-based conception of solidarity. We will begin by reviewing in more detail the earlier work of Markovsky and Lawler (1994). Following this, we will argue the benefits of incorporating social identification as a basis for solidarity and then explore the implications of various types of external bond-breaking forces.

Previous Work

Markovsky and Lawler (1994) provided a theory containing a purely structural definition for solidarity. As noted earlier, it stated that "a solidary group is a set of actors with high reachability and unity of structure" (p. 123). *Reachability* is a concept from the social network literature related to centrality. Most often this concept is treated as a property of positions in networks. Markovsky and Lawler treated it as a network-wide property: a network has high reachability to the extent that all of its actors are connected to one another through strong and direct attractive bonds. Reachability is diminished to the degree that bonds are weaker, less attractive or repulsive, or that actors tend to be connected to others only through indirect relations.

Unity of structure referred to the relative absence of substructures within a set of actors, that is, structural homogeneity. All else being equal, homogeneous networks are by definition more solidary than "clumpy" or clique-filled networks. Importantly, homogeneity and *density* need not covary.

For example, a relatively "open" or sparse structure with few relations but no hierarchy or clique structure may have a higher degree of solidarity than a much more densely connected heterogeneous network.

Markovsky and Lawler also provided the following definitions:

- *Cohesion:* A set of actors is cohesive to the degree that it has high reachability.
- *Aggregate:* An aggregate is a set of actors with low cohesion.
- *Assemblage:* An assemblage is a set of actors with intermittent cohesion.
- *Solidarity:* A solidary group is a set of actors with high reachability and unity of structure.

In formulating these definitions, Markovsky and Lawler were inspired by analogies to phase transitions in states of matter—gas, liquid, solid—and various abstract properties that they share with different types of social groupings. Doing so helped to clarify how micro-relations in networks may generate emergent properties at the group level. Corresponding to gaseous states, the concept of *aggregate* was defined in a way typifying its informal usage in social psychology. More so than aggregates, *assemblages* entail interpersonal ties of limited strength and/or duration, in essence creating "liquid" groupings of actors. Corresponding to solid states of matter, *solidary groups* are characterized by their relative stability. Actors connected via "impure" mixed relations may exhibit intermediate forms with properties analogous to "slushy" or "rubbery." In both material and social aggregates, such emergent properties emanate under certain conditions involving the valences and strengths of bonds that link constituents, and the patterns formed by those related members. These patterns may be characterized by such adjectives as dense, wispy, flat, layered, crystalline, hierarchical, centralized, diffuse, homogeneous, lumpy, and so on.

In addition to these group attributes, the Markovsky-Lawler (M-L) conceptualization of solidarity generated a number of other implications. For instance, among the group-level properties associated with different degrees of solidarity were such qualities as degree of structural strength, the level of the largest discernible and enduring subunit, the level at which chaotic activity occurs, and the clarity and penetrability of the group boundary. The concept was further exemplified via a theory of emotion-based group solidarity. It was shown that solidarity would emerge with the satisfaction of a small set of assumptions pertaining to emotional experiences and interpersonal actions.

From a sociological standpoint, there were many advantages to this structural definition. First, it transcended a quagmire of loosely formulated ideas and theory fragments. There is a great deal of inconsistency in definitions and operationalizations of solidarity and cohesion terminology, and virtually no recognition in the relevant literatures that this is even a minor problem, much less a fatal one. By failing to make definitions and theoretical propositions explicit, all possibility is lost of employing the most powerful tool of science: the collective evaluation of theoretical claims. Theorists have no assurance that others grasp the intended meaning of their claims, and researchers have no guidance—beyond their personal intuitions—in operationalizing theoretical concepts in their tests and applications. This accounts for the disheveled state of the cohesion literature.

Second, the M-L definition is more explicit than prior conceptualizations of solidarity with respect to levels of analysis. Although it is common to find theorists speaking of solidarity and cohesion as properties of groups, the concepts are invariably treated as sets of individual-level dispositions about other individuals, or forces acting upon individuals. Absent is theory and research on the role of higher-level patterns of relations—patterns that simply cannot exist at the level of individuals or of dyadic relations.

As a simple example of the consequences of considering higher-level patterns, suppose we know that each member of a set of eight people is socially (and mutually) attracted to two others and is indifferent toward five. What can we say about the solidarity of these people compared to, say, a set of eight people in which each is attracted to three and neutral toward four? The first pattern could be analogous to a subset of a community in which families are interconnected through children or places of work, but there is no real "center" to the network and most network members have ties to relatively few others. The second pattern is more akin to ethnic enclaves within a community in which interpersonal ties are strong within, but weak between the subgroups.

Figure 1 shows two different configurations for relationships in the two hypothetical sets of actors. In accord with the conditions given here, Network 1 contains a structure in which each of the eight actors is socially attracted to exactly two others and indifferent toward five. In Network 2, each person is attracted to three others and indifferent toward four. This is drawn as two sets of actors with each-to-all relations within sets and no relations between sets. Network 1 has both unity of structure (no substructures) and, though many relations are indirect, each can reach all others. In Network 2, each position has four actors that it cannot reach and there are two very pronounced substructures. By the M-L definition, Network 2 has lower solidarity. The set of actors it portrays "falls apart" into two subsets. Actors in Network 1 form a single, more integrated structure. Importantly, cohesion theorists, with no real distinction for "cohesion" and "solidarity," would have to say that cohesion is lower in Network 1 because it has only two-thirds the number of attractive

Network 1

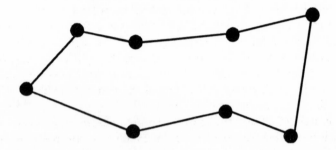

Network 2

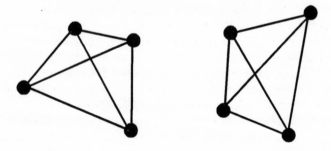

Figure 1. Structured Relations in Two Networks

bonds as appear in Network 2. Shifting perspective to the group structural level provides a very different set of implications.

The abstractness of the definition of solidarity has further advantages. By making solidarity a property of *sets of actors*, it can subsume both familiar types of social groupings involving individual human members—formal organizations, families, and so forth—or unfamiliar groupings containing either individual or corporate actors, the theoretical utility of which remains to be discovered. Either way, the definition permits us to build multilevel theories that link lower-level with higher-level units (and, possibly, with even lower or even higher units) and apply the same structural principles in describing the solidarity of resulting structures. Solidarity becomes a conceptual tool that allows us to look at a plethora of group phenomenon that occur within large groups with embedded groups, allowing us to speak to the solidarity at different levels of organization and structure as well as the overall solidarity of the broader network.

A structural definition of solidarity also provides an improved focal point for explicit theorizing. As noted earlier in several contexts, previous work has generally failed to distinguish between the causes, consequences, and defining properties of solidarity and cohesion. As a result, pointless arguments ensue among those who want to declare what "cohesion" *really* means, or which of its determinants are most important in some universal sense. Viewed as a whole, the research has reflected this confusion by its pronounced emphasis on finding statistical relationships among empirical indicators, with little or no attention paid to such questions as whether those indicators have measured a defining property, a cause, or a consequence of solidarity.

Suppose, for example, that in a particular study a high correlation is found between a cohesion measure and an interpersonal attraction measure. Such a finding is only meaningful when interpersonal attraction is not among the defining properties of cohesion within the theory presumably being tested by the research. Without the existence of a prior theory that forces certain unambiguous hypotheses about empirical relationships, such a finding may easily be interpreted ex post facto to support what is actually an ad hoc or even post hoc hypothesis.

The M-L conceptualization does have several limitations. First, it would benefit by a more formalized mode of expression. In particular, it did not provide explicit mathematical indices for reachability and unity of structure. Second, it did not include social identification as a basis for solidarity. Third, the M-L definition paid no special attention to forces that may weaken particular social ties and diminish group solidarity. We believe that specific types of forces will have more or less of an impact on a group's solidarity, depending on the particular manner in which the group's solidarity is structurally manifested. Further, once solidarity is broken, the remaining substructures may be expected to vary in their degrees of solidarity depending upon various internal and external contingencies. In the remainder of this chapter we will focus on the latter issues and leave problems of formalization to future work.

Social Identification as a Basis for Solidarity

Social identification provides a type of attractive bond through which the solidarity of a network may be established or enhanced. We believe that opening up the solidarity *concept* to such non-interpersonal bonds adds a powerful new component to a solidarity *theory* employing this definition. Doing so provides dual bases for the maintenance of group structures: first, the patterns of attractive interpersonal bonds and, second, the strength of social identity bonds between persons and their cognitive representations of the group.

Graphically, we suggest a provisional method for representing an identification bond: a relation drawn between an actor and a node, where the node is a symbolic representation of the set of actors (or group) with which the actor identifies. In keeping with the existing formulation, two or more actors that identify with the same group are mutually reachable by at most two steps in the network. Elaborating the earlier example of the divided network Network 2, Figure 2 shows how adding identification bonds that do not strictly correspond to the existing pattern of attraction bonds enhances reachability and weakens the prominence of the substructures. In this case, the pattern of social identifications would enhance the solidarity of the 8-actor set.

The figure illustrates one method of formalizing the notion from social identity theory that when individuals link themselves (literally, *their selves*) to the group via self-identification, they strengthen their links to one another. The minimal-group literature suggests that the perception of commonality, whether around goals, ideologies, places of origin, or shared fate, triggers group-oriented behaviors. This perception-cognition-behavior sequence suggests that reachability as established through indirect identification ties (as in Figure 2) probably facilitates the development of more direct ties among actors with common identifications.

This line of thinking also reconnects to, and subsumes, Durkheim's view on mechanical solidarity. Actor A does not have to be in the presence of B, or even aware that B shares the same self-categorization, to be attracted to or act in reference to B. And yet, such ephemeral ties facilitate the triggering of more direct ties given some salient event. When a U.S. pilot is shot down

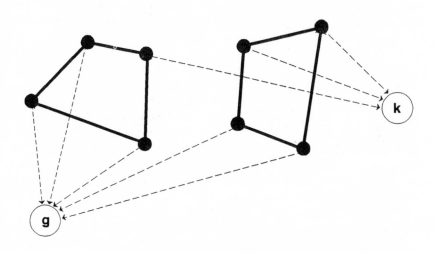

Figure 2.   Identification Bonds

and dragged through the streets of Mogadishu, Somalia, thousands upon thousands of Americans called the White House to express support for this fellow group member, and anger at the out-group, though virtually all callers shared nothing with this unfortunate pilot beyond their group identification.

What factors weaken solidarity? Implicitly, the M-L definition assumed that if any stress is placed on relations in a network, it occurs either uniformly or randomly. In actuality, however, there are many ways that the solidarity of a group—or solidity of a physical object—may be broken. All solidarity-breaking processes depend on the weakening or severing of at least some of the bonds that maintain the network structure. The forces that accomplish this breakage may be large and diffuse relative to the structure. In the physical case, a heavy snow collapses a roof; in the social case, all of the members of a once close circle of friends are gradually drawn away by other interests. The forces may also be small and surgical, as when a chain is broken by the removal of a link, or a rumor-mill is dissolved by the departure of a central gossip. In Figure 3, if forces were applied such that only the three indicated social identification bonds were broken, the level of solidarity would be reduced practically to where it was before the identity bonds were added.

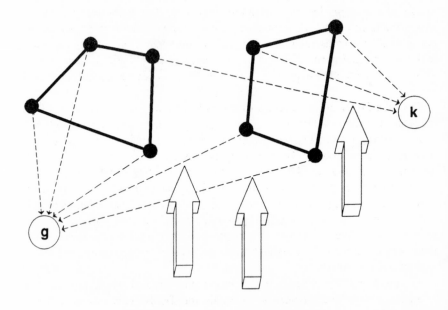

Figure 3.   Bond-breaking Forces

Disintegrative Forces

Different structures may be solidary in different ways and to different degrees. These differences have implications for the ability of those structures to resist the disintegrating effects of bond-breaking forces. For example, the solidarity of a typical nuclear family is greatly enhanced by its multiplexity—the existence of multiple types of relations among members. Each of its members is connected to every other by a variety of relations that may include physical dependence, emotional dependence, friendship, companionate love, maternal love, and so forth. Members of a Tuesday night poker group may also have each-to-all relations, but the variety or strength of those relations may be very minimal. Both groups have a degree of solidarity, but the lure of alternative sources of Tuesday evening recreation should have a far more disintegrative effect on the poker group than on the family. We can say that all else being equal, greater multiplexity means lower vulnerability to disintegrative forces.

On the other hand, all else is not equal and bond-breaking forces may vary along at least four dimensions: *source, breadth, strength, and specificity.* First, the *source* of disintegrative forces (d-forces) may be either internal or external to a set of actors. Sibling rivalries in families can weaken attractive bonds from within the family network, or a civic group's lack of activity may weaken the group's salience to its members and, as a result, their sense of identification with the group. External d-forces are also common, as when one group lures away another's members or tries to weaken the collective capacities of a rival group.

The *breadth* of d-forces refers to the proportion of relations upon which they act. To bust a labor union, it may be necessary to sway the allegiance of a majority of members by weakening their interpersonal bonds as well as their identification bonds. To bust an evangelical sect bound primarily by social identification, it may only be necessary to scandalize the leader.

The *strength* of d-forces refers to their ability to weaken bonds. Making a new friend may slightly weaken relations with one or more other friends— if it demands personal resources that would otherwise be spent on those other relations. From the perspective of its employees, selling off a corporate division is a strong and drastic d-force that breaks numerous communication, exchange, and informal network ties.

Finally, *specificity*. D-forces may or may not be directed at specific structural relations. An interesting example of highly specific d-forces was the attempt by the FBI in the 1960s to weaken and dissolve "New Left" and civil rights groups. Programs were developed to prevent the rise of charismatic leaders, restrict the activities of individuals deemed trouble-makers, and quash members' identification with groups by publicizing false statements supposedly made by prominent group members. An example of a diffuse, nondirected d-force could be the tug of nonwork-related commitments that most of us experience at any given time. These forces are not tailored to systematically weaken our academic department's solidarity, but we suspect that they invariably do so, to at least some degree.

To speak of structural disintegration demands that post-disintegrated structures be addressed. Our approach to this issue is a bit unusual, relatively simple and, we think, quite powerful. First, we suggest that the concept of solidarity should be thought of as not necessarily pertaining to existing groups of actors but rather to *sets* of actors defined as such for any analytic purpose. The set may be a tightly connected group of intimates or it may be a random selection of members of a community. We would want to be able to address the solidarity of either set.

For example, assume that we have selected or assembled a set of ten actors connected each-to-all, as shown in Figure 4a. To this high-solidarity ten-actor set is applied a series of strong and highly surgical d-forces such that what ultimately remain are two completely separate five-actor subsets with each-to-all ties within each subset (Figure 4b). The solidarity of the original set is now considerably weakened because both the reachability and structural unity criteria are breached. On the other hand, nothing prevents us from addressing relative degrees of solidarity of the individual subsets. Each has high solidarity. In fact, had our original analytic focus been on what ultimately became a subset of five, we would have declared that no change in the solidarity of that set took place after the application of the d-forces. Thus, the effect of d-forces depends upon how we choose to delimit a set of actors for theoretical or analytic purposes.

An analogy can be drawn to solid physical objects. If you hold in your hand a steel bolt with a nut screwed on tightly, you would say that it is a relatively solid object. True, it has identifiable components, but they are firmly integrated and fixed with respect to one another. You may then apply a relatively weak, but narrow and specific force to break the solidity of the object: you unscrew and remove the nut. Holding these pieces in your hand, you no longer hold a relatively solid object. At the same time, had your original interest been just on the bolt, you would have declared its solidity to be unaffected by the removal of the nut. In social terms, the effect of the disintegrative force only becomes meaningful when it changes relations in an analytically delimited network.

This broadened conceptualization of solidarity and solidarity-breaking forces has its theoretical costs and benefits. By way of costs, it places greater empirical demands upon the investigator or practitioner: to measure all of the potential sources of solidarity of a given set of actors, it is first necessary to account for the number and strength or salience of all types of non-neutral relations among members of the set. Then the level of d-forces at a given point in time must be determined, and the location and type of relations to which they are relevant. However, the fact that certain structural forms of natural groups are relatively common further eases the empirical burden. The benefit of all this, of course, is that one obtains a much more accurate picture of the structure of relations among members. This, in turn, shows the degree to which d-forces of a given type will affect the reachability in, and unity of, the structure.

Following the earlier example of multiplex relations in families, it was stated that all else being equal, greater multiplexity means greater invulnerability to disintegrative forces. For any given set of actors with multiplex ties, however, it becomes essential to determine the patterns formed by the combination of the different relational bases. Multiple ties between a given pair of network members often form in a self-reinforcing manner, as when acquaintances become friends, and these friends engage in joint activities, become dependent on one another, communicate frequently, and so on. By itself, however, such a trend may foster dyadic solidarity, but does nothing for any larger set in which the dyad may be imbedded.

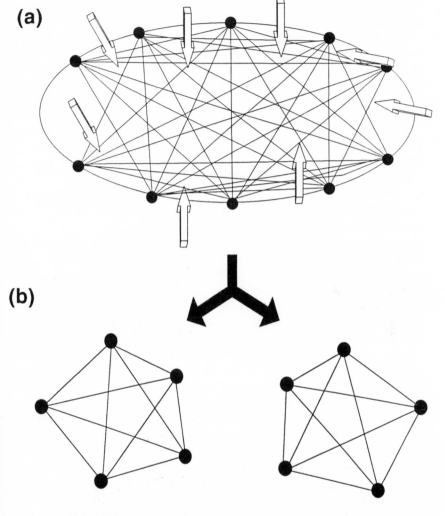

(a)

(b)

Figure 4. Network Disintegration, Sub-network Solidarity

On the other hand, to the degree that the different relational bases in a multiplex network form different but interlocking patterns, rather than insular subsets of actors, solidarity is by definition enhanced. To the degree that subsets of actors are interlocked by only tenuous relations—low strength and/or low multiplexity—then the larger set is vulnerable to the degree that prevailing disintegrative forces may act upon those specific types of relations. In short, the generalization from uniplex to multiplex networks seems relatively straightforward.

CONCLUSIONS

As for the future of this line of work, we see a clear need for a more explicit solidarity metric, especially with the consideration of solidarity in multiplex networks. This is the most important task now facing us. To capture the two defining properties of solidarity, the measure will have to account for two corresponding network properties. First, we must either create or choose a method for codifying and aggregating the levels of reachability among members of a set of actors. This means specifying values for disconnection, one-step connection, two-step connection, and so forth, and specifying how connections of varying strength combine.

Second, a cluster- or block-detecting algorithm will be needed to capture the degree of structural disunity or heterogeneity. Available network analysis programs contain cluster analysis and blockmodeling modules, and we are presently considering their usefulness for our purposes.[3] It is highly doubtful that there will be a single omnibus heterogeneity index suitable as an empirical measure of solidarity. Instead, specific indices within a more general class will probably have to be chosen or engineered for specific purposes.

Third, we will have to provide a synthetic mathematical expression that in some reasonable and useful way combines these reachability and heterogeneity indices. We may find that for different theoretical purposes, one or the other criterion should be more heavily weighted through parameterization.

Fourth, it makes little sense to develop the concept and measure of solidarity in the absence of a theory that makes use of it. Concepts are not testable in and of themselves; only statements asserting something about how they relate to other concepts can be tested. This is not a serious problem. We have shown, as did Markovsky and Lawler (1994), that this concept of solidarity is very amenable to causal statements of the sort offered in past formulations, and to a variety of new possibilities such as emotion-based and identity-based group solidarities.

Finally, we are especially optimistic about the use of computer simulations. With them we can study networks of any structural configuration, either realistic or idealized, and having any distribution of ties or degree of multiplexity. We can inflict upon them d-forces that vary along the four dimensions mentioned earlier and explore in a systematic way how different relational patterns fare under different patterns of disintegrative forces. While all this is happening, integrative forces may operate and countervail the bond-breaking forces. In short, there seems to be plenty here to keep us busy.

ACKNOWLEDGMENT

We would like to thank the following people for their useful comments on earlier versions of this paper: Karen Heimer, Ed Lawler, Mark Mizruchi, Jodi O'Brien, and Bob Shelly.

NOTES

1. Within a given theory, definitions associate explicit sets of properties with particular theoretical terms. For a given term, its listing of properties allows the researcher to classify any empirical entity as either an instance (or "operationalization") of the concept or as a non-instance. Each term in the definition must be either previously defined in the theory, or else "primitive," that is, its meaning is presumed to be already known to members of the theory's intended audience. When a theorist conflates defining properties and determinants (i.e., makes tautologous statements), he or she renders that theory unfalsifiable, hence untestable.

2. Hechter saw one of the problematic threads in solidarity analysis emerging from the mechanical-organic, Gemeinschaft-Gesellschaft distinction. He felt theorists presumed that there had to be at least two different theories to explain the differences in these groups. Normativist analysis offered that Gemeinschaft was the result of internalized norms. Most problematic with that account was the inability to distinguish a normatively induced action from one performed out of fear of or desire for a sanction. Hechter saw this dilemma as reducing normative explanations to a tautology.

3. For example, STRUCTURE, published by The Center for the Social Sciences at Columbia University, and UCINET IV from Analytic Technologies contain a variety of measures of structural cohesion and clique detection that may prove useful for such purposes.

REFERENCES

Bonner, H. 1959. *Group Dynamics: Principles and Applications.* New York: Ronald Press.
Cartwright, D. and A. Zander. 1968. *Group Dynamics: Research and Theory.* New York: Harper and Row.
Durkheim, E. 1956/1893. *The Division of Labor in Society.* New York: Free Press.
Eisman, B. 1957. "Some Operational Measures of Cohesiveness and their Interrelations." *Human Relations* 12: 183-189.
Festinger, L., H. Reicken, and S. Schachter. 1956. *When Prophecy Fails.* Minneapolis: University of Minnesota.
Festinger, L., S. Schachter, and K. Back. 1950. *Social Pressures in Informal Groups.* New York: Harper Bros.
Gamson, W. A., B. Fireman, and S. Rytina. 1982. *Encounters With Unjust Authority.* Chicago: Dorsey.

Gerard, H.B. and M.F. Hoyt. 1974. "Distinctiveness of Social Categorization and Attitude Toward Ingroup Members." *Journal of Personality and Social Psychology* 29: 836-842.

Hechter, M. 1987. *Principles of Group Solidarity*. Berkeley: University of California Press.

Heider, F. 1958. *The Psychology of Interpersonal Relations*. New York: Wiley.

Hogg, M. A. 1985. "Masculine and Feminine Speech in Dyads and Groups: A Study of Speech Style and Gender Salience." *Journal of Language and Social Psychology* 4: 99-112.

———. 1987. "Social Identity and Group Cohesiveness." In *Rediscovering the Social Group: A Self-categorization Theory*, edited by J. C. Turner, M. A. Hogg, P. J. Oakes, S. D. Reicher, and M. Wetherell. Oxford: Blackwell.

Hogg, M. A. and D. Abrams. 1988. *Social Identifications: A Social Psychology of Intergroup Relations and Group Processes*. London: Routledge.

Jackson, J. M. 1959. "Reference Group Processes in a Formal Organization." *Sociometry* 22: 307-327.

Latané, B. 1981. "The Psychology of Social Impact." *American Psychologist* 36: 343-356.

Lewin, K. 1948. *Resolving Social Conflicts*. New York: Harper and Bros.

———. 1952. *Field Theory in Social Science*. London: Tavistock.

Lott, A. J. and B. E. Lott. 1965. "Group Cohesiveness as Interpersonal Attraction: A Review of Relationships with Antecedent and Consequent Variables." *Psychological Bulletin* 64: 259-309.

Markovsky, B. and E. J. Lawler. 1994. "A New Theory of Group Solidarity." In *Advances in Group Processes* (Volume 11), edited by B. Markovsky, K. Heimer, and J. O'Brien. Greenwich, CT: Jai Press.

McGrath, J. E. and D. A. Kravitz. 1982. "Group Research." *Annual Review of Psychology* 33: 195-230.

Nagel, E. 1979. *The Structure of Science*. Indianapolis: Hackett.

Paulus, P. B. 1983. "Group Influence on Individual Task Performance." In *Basic Group Processes*, edited by P. B. Paulus. New York: Springer Verlag.

Porter, L. W. and E. E. Lawler. 1968. *Managerial Attitudes and Performance*. Homewood, IL: Richard D.Irwin.

Ramus-Nienhuis, W. and A. van Bergen. 1960. "Relations Between Some Components of Attraction-to-group: A Replication." *Human Relations* 13: 271-277.

Shaw, M. E. 1981. *Group Dynamics: The Psychology of Small Group Behavior* (3rd edition). New York: McGraw-Hill.

Tajfel, H. 1972. "Social Categorization." English translation of "La Categorisation Sociale." In *Introduction a la Psychologie Sociale* (Volume 1), edited by S. Moscovici. Paris: Larousse.

Turner, J. C. 1975. "Social Comparison and Social Identity: Some Prospects for Intergroup Behavior." *European Journal of Social Psychology* 5: 5-34.

———. 1982. "Towards a Cognitive Redefinition of the Social Group." In *Social Identity and Intergroup Relations*, edited by H. Tajfel. Cambridge: Cambridge University Press.

———. 1984. "Social Identification and Psychological Group Formation." In *The Social Dimension: European Developments in Social Psychology* (Volume 2), edited by H. Tajfel. Cambridge: Cambridge University Press.

Turner, J. C., M. A. Hogg, P. J. Oakes, S. D. Reicher, and M. Wetherell. 1987. *Rediscovering the Social Group: A Self-Categorization Theory*. Oxford: Blackwell.

Turner, J. C. and M. A. Hogg, P. J. Turner and P. M. Smith. 1984. "Failure and Defeat as Determinants of Group Cohesiveness." *British Journal of Social Psychology* 23: 97-111.

Turner, J. C., I. Sachdev, and M. A. Hogg. 1983. "Social Categorization, Interpersonal Attraction and Group Formation." *British Journal of Social Psychology* 22: 227-239.

Zander, A. 1979. "The Psychology of Group Process." *Annual Review of Psychology* 30: 417-451.

MUTUAL DEPENDENCE AND GIFT GIVING IN EXCHANGE RELATIONS

Edward J. Lawler, Jeongkoo Yoon, Mouraine R. Baker, and Michael D. Large

ABSTRACT

We incorporate elements of a social-constructionist viewpoint into social-exchange theory and show how mutual dependence can produce expressive behavior in the form of gift giving. Exchange networks typically create varying degrees of mutual dependence in component dyads, and greater mutual dependence produces more frequent exchange. We propose that over time, frequent exchange generates an expressive relation and unilateral, token gifts are an indicator of emerging expressiveness in an exchange relation. To experimentally test the impact of mutual dependence on token gift giving, two focal actors, each with one alternative partner, attempted to negotiate an exchange across multiple opportunities. The results indicate that high compared to low mutual dependence increased gift giving, while also enhancing the attitudinal commitment to and perceived closeness of the relation. Consistent with the theory, these effects of mutual dependence were indirect, operating through the frequency of exchange. Broadly, the paper has theoretical implications for how and when endogenous processes in dyads generate certain micro-to-macro effects.

Advances in Group Processes, Volume 12, pages 271-298.
Copyright © 1995 by JAI Press Inc.
All rights of reproduction in any form reserved.
ISBN: 1-55938-872-2

Exchange theory adopts the instrumental premise that actors enter and remain in exchange relations to the extent that these provide more benefit than likely from alternative relations (Homans 1961; Blau 1964; Emerson 1981). Power dependence theory portrays the instrumental foundation as a set of positions and the relations among them—that is, a social structure—in which actors pursue their individual interests through interaction with others (Cook and Whitmeyer 1992). To the degree that a structure leads the same actors to exchange repeatedly over time, an interpersonal relation is likely to form just as groups tend to form around collective experiences and a sense of mutual fate (Rabbie, Schot, and Visser 1989; Kramer 1991; Lawler and Yoon 1993). Such a process has important implications for exchange networks and raises the possibility of incorporating select features of social-constructionism into social-exchange accounts for micro social orders.

Broadly, this paper conceptualizes the development of an exchange relation in a network as incipient "group formation".[1] The primary focus is how and when *mutual* dependence in an exchange relation produces commitment behavior in the form of token gifts. *Mutual dependence* is the total or average power in an exchange relation (Emerson 1972, 1981; Bacharach and Lawler 1981; Lawler 1992a; Molm 1987, 1990), that is, the average difference between prospective payoffs within versus outside the focal relation or CLalt (Thibaut and Kelley 1959). Emerson (1972, 1981) termed this "relational cohesion," thus implying group-formation effects for mutual dependence. *Token gifts* are small, unilateral, unconditional benefits that have the shared meaning that actors are willing to "give without expecting anything in return." Gift giving is particularly relevant to exchange relations because under some conditions it suggests the transformation of a purely instrumental relation into one with expressive components. We interpret such acts as a rudimentary form of everyday ritual behavior, symbolic or expressive of common membership in a relation, group, or organization (see Wuthnow 1987; Collins 1981, 1989 for relevant discussions of ritual behavior).

To understand how structural power might foster ritual behavior, in general, and gift giving in particular, this paper builds from common themes of literatures on social exchange (Emerson 1972, 1981), the social construction of reality (Berger and Luckmann 1966), and ritual as an everyday phenomenon (Collins 1981, 1989; Turner 1982; Wuthnow 1987). Our general theoretical argument is that mutual dependence in an exchange relation (dyad) is likely to produce some forms of ritual behavior after an initial period of instrumental exchange; and gift giving is a form of special import to exchange theory. There is a long tradition in exchange theories of considering gifts as a distinct form of exchange constituting a moral statement or definition about the relationship of self and others or self to a larger group (Heath 1976; Ekeh 1974; Arrow 1972).[2]

In a recent study, Lawler and Yoon (1993) examined whether an emotional/affective process mediates the impact of equal versus unequal power dependence on gift giving. They treated gift giving as one of two behavioral indicators of commitment in negotiated exchange, the other being stay behavior, that is, remaining in a relation despite equal or better alternatives. The main argument of their theory was that frequent exchange between the same actors is joint behavior that mediates the impact of structural power on gift giving, ostensibly because repeated exchange arouses positive feeling or emotion (Lawler and Yoon 1993). Consistent with the theory, equal compared to unequal power produced more commitment behavior indirectly by enhancing the frequency of exchange and by producing more positive emotion. Lawler and Yoon's (1993) broader interpretation was that frequent exchange makes the exchange relation a social object, and members of the dyad become affectively attached to that relation because it is perceived as a source of the positive emotion. In this way, equal power ostensibly produces commitment behavior through an emotional/affective process.

Considering token gift giving as a form of everyday ritual leads to some important theoretical implications that we elaborate in this paper. If gifts are small token items of little instrumental value, it seems reasonable to treat them as representing a "ritualized" type of commitment behavior. Treating token gifts as ritual does not imply that gifts are the only type of ritual or that other forms of commitment (i.e., stay behavior) cannot become ritualized, but it does make gift giving a special form of commitment behavior and enables us to graft social constructionist ideas about ritual (Berger and Luckmann 1966) onto a social-exchange framework.

Developing and forging links between exchange theory and social constructionism is complicated by the fact that these traditions contain disparate assumptions about actors and social systems and also about how to understand or explain social phenomena. We adopt the ontological and epistemological approach of exchange theory and suggest that overlapping ideas and themes—reflected in concepts of commitment, institutionalization, and ritual—provide the basis for building elements of social constructionism into social-exchange theory. Our theory, as a variant of social exchange, assumes actors with instrumental ties who interact and can develop the sort of emotional bonds that social constructionists and Durkheimians typically assume to be present at the outset of the interaction (Durkheim 1915; Collins 1981; Berger and Luckmann 1966).

BACKGROUND

The following pages elaborate key concepts of ritual and gift giving, and theorize a link between the "objectification of an exchange relation" and token gifts as everyday ritual behavior. We then formulate several hypotheses about the emergence of gift giving in negotiated exchange and test them experimentally.

Concept of Ritual

The concept of "ritual" has historically been associated with collective experiences that affirm for members their common identity, culture, history, or future (Durkheim 1915; Malinowski 1922; Levi-Strauss 1969). The prototype is often taken to be religious ceremonies in which actors engage in highly focused and emotional behavior with symbolic meaning. This is a fairly limited conceptualization (see Goffman 1967; Collins 1975). Ritual can be construed more broadly as any expressive behavior, undertaken for its own sake and symbolic of a tie to larger collective—a relation, group, or organization (Collins 1975). Goffman (1967) showed that such behaviors are a normal, everyday phenomenon by which actors affirm and shape their ongoing relations with other actors.

To Goffman (1967), everyday ritual behavior communicates common purpose, shared definitions of self and other, and trustworthiness. He emphasized the communicative value of ritual in social interaction, treating it as symbolic behavior with a subtle, but clear, practical function (see also Turner 1982). Others, such as Berger and Luckmann (1966) emphasize how ritual behavior ties individuals to a collective (see also Collins 1975; Wuthnow 1987), which is more relevant to our concerns. Token gift giving, we suggest, reflects an emotional/affective actor-to-collective tie (Lawler and Yoon 1993) and presupposes that actors perceive their relation as an objective unit.

Berger and Luckmann's (1966, pp. 53-60) view is that the *objectification* of a collective entity underlies ritual behavior. They suggest two basic conditions for objectification and ritualized behavior: (1) the "habitualization" of action, such as, repetitive behavior; and (2) "reciprocal typification" or shared definitions of that repetitive behavior. We apply this view to emerging exchange relations. Habitualization is treated as a basis for typification, and both of these processes are the basis for objectification. Actions are ritualized once they are defined by actors as "This is how things are done [here]" (Berger and Luckmann 1966, p. 59). Objectification of an exchange relation presupposes "incipient institutionalization" of a line of behavior.[3]

Importantly, Berger and Luckmann (1966) argue that "all actions repeated once or more tend to be habitualized to some degree" (p. 57), which starts the objectification process. This condition, habitualization, is an inherent characteristic of an "exchange relation," defined by Emerson (1981, p. 42) as a "series of transaction by the same actors over time." Both Berger and Luckmann's social-constructionist and Emerson's social exchange framework imply that the emergence of the social—an incipient *social* order—stems from repetitive behavior when the same people interact. A similar theme is echoed by Collins' (1975, 1981, 1989) theory of interaction ritual chains, which ascribes the source of ritual to emotional energy generated by repeated "conversations" among the same actors. In fact, the essential difference between social and economic exchange, according to Emerson (1981), is that the interconnections between transactions among the same actors produce social effects such as trust and commitment. These effects are social constructions in Berger and Luckmann's sense, social constructions that reflect objectification and incipient group formation (Callero 1991; Kollock and O'Brien 1992). We contend that objectification stems from "habitualization" and are likely to be manifest in ritualized gift giving in negotiated exchange.

The proposition implied by this reasoning is quite simple: *Habitualization of interaction leads to objectification of a larger social unit* [*e.g., an exchange relation*] *which, in turn, produces ritual behavior.*[4] It would be easy to confuse the repetition of exchange with ritual, because ritual is so closely associated with habit. However, repeated exchange is not sufficient to reflect the symbolic or expressive content of ritual; something more than habit or repetition is needed to make the behavior symbolic of a person-to-collective tie and relevant beyond the immediate situation. Ritual behavior has significance, beyond the completion of the immediate act to the degree that it symbolizes or is expressive of the relations between the individuals or group membership (Turner 1982; Callero 1991). In exchange networks, we suggest that token gifts can be interpreted as such symbolic expressions (see also Heath 1976).

Concept of Gift Giving

Gift giving is a complex, multifarious phenomenon, as is evident from the variety of conceptions and approaches in the literature (Mauss 1954; Schwartz 1967; Titmuss 1971; Arrow 1972; Ekeh 1974; Heath 1976; Akerlof 1982, 1984; Haas and Deseran 1981; Caplow 1982, 1984; Cheal 1986, 1988; Jasso 1993). Exchange approaches generally stress that gifts must be reciprocated or they will cease to occur (Emerson 1972; Blau 1977; Caplow 1984; Cheal 1988); they look for evidence of implicit, subsurface, long-term reciprocity as an explanation for gift giving (Gouldner 1960). Such approaches also stress that gifts must be costly to the giver to impress the other; some even contend that the costs to the giver must be greater than the benefits to the receiver (Schwartz 1967), otherwise what appears to be gift giving is simply another way for actors to generate joint benefit. Overall, exchange theorists absorb gift giving into a broader utilitarian web and attempt to capture how gifts actually enhance the benefits actors receive from their relationship (Emerson 1981; Heath 1976; Akerlof 1982, 1984).

Social constructionists view gifts as symbolic communications, putting forth a definition of a relationship (e.g., Haas and Deseran 1981), an identity (Schwartz 1967), or a sense of community (Arrow 1972). Gifts by A communicate to B that A defines the interpersonal relation as a trusting, friendly one—and some form of reciprocal action by B confirms that this is a shared definition. In a social constructionist account, reciprocity remains central over time, but the reciprocation of benefits is not important in itself. Of most importance is evidence of reciprocal, convergent definitions of the relation. From this standpoint, the utilitarian value of gifts is relatively trivial.

Our focus is fairly specific and narrow in this paper. We are concerned with the rate of initiating gifts, not reciprocation of them, in a dyadic relation. There is no requirement to give and only a slight loss in giving gifts. This is a very rudimentary form of gifts that is designed to reflect whether actors are beginning to treat the exchange relation as a positive social object toward which expressive acts are directed. Extrapolating from both exchange and social constructionist viewpoints, gifts in an exchange relation can be interpreted as ritual behavior if they are *token* (i.e., have little or no extrinsic value), *unilateral* (i.e., carry the connotation that actors are giving without knowing whether the other will), and *noncontingent* (i.e., there is no explicit expectation of reciprocity). It is also important that gifts come from a value domain distinct from those at issue in the exchange (Emerson 1981). The gifts of concern are small, minor items with primarily symbolic import.

HYPOTHESES

We argue that gift giving is the result of an endogenous process through which repetitive exchange produces a definition of the relation as a unit. Two sets of hypotheses integrate basic ideas from social exchange and social constructionist viewpoints. The first set, based on exchange theory (Emerson 1972; Bacharach and Lawler 1981; Molm 1987; Lawler 1992b), predicts effects for mutual dependence. The second set, based on both exchange and social constructionist theories, predicts effects for uncertainty about the payoffs from alternative relations. Our hypotheses propose that power-dependence relations and uncertainty about an alternative relation promote the social construction of an interpersonal relationship by producing more frequent exchange in a focal dyad. Frequent exchange sets apart some exchange opportunities or dyadic relations from others, helping to make the focal relation a distinct "social object" (Lawler and Yoon 1993).

Mutual Dependence Hypotheses

Emerson's (1972) power dependence theory conceptualizes power as the structural potential to influence and distinguishes this potential from power use or actual, realized power (see also Cook and Emerson 1978; Bacharach and Lawler 1981; Lawler 1992a; Molm 1987, 1990). In an exchange relation, A's structural potential is a function of B's dependence on A, and B's structural potential is a function of A's dependence on B. Each actor's dependence is in turn a positive function of the value of benefits received in exchange and an inverse function of the availability of such benefits from alternative relations. As with most empirical studies of power dependence, we hold constant the value at stake in the focal relation and examine variations in dependence based on the nature of the alternatives available.

We build on the implicit nonzero sum concept of power in power dependence theory (Lawler 1992a; Lawler and Ford 1993). A nonzero sum conception allows for variation in the total amount of power potential in a relation (i.e., mutual dependence), while a zero sum conception assumes that the total amount of power in a relation is fixed. The nonzero sum conception in power dependence is manifest, for example, in the fact that changing networks could give both actors in a dyad more alternatives resulting in each being less dependent on the other, or fewer alternatives making them more dependent. Mutual dependence can vary independently of relative dependence (see Lawler 1992a for more discussion).

Emerson (1972) characterized changes in mutual dependence as changes in "relational cohesion," or the structural push toward collaborative action. Relational cohesion is simply the difference in expected value of exchange in the focal dyad versus exchange in an alternative one (CLalt), averaged across actors in a dyad (Molm 1987; Lawler 1992a). In this study, we ask how variations in total or mutual power dependence—given equal dependence— affect exchange frequency and the initiation of token gifts.

The mutual dependence of actors in negotiated exchange captures the susceptibility of the exchange relation to disruption. The larger the difference between the possible benefit within the focal relation and the alternative, the less vulnerable the focal exchange relation because, among other things, there is more room for misjudgment or miscalculation. Higher total power or mutual dependence means a larger number and range of negotiated agreements can meet a "sufficiency criterion"—that is, provide more benefits than likely from the alternative. An exchange relation with higher mutual dependence provides the flexibility to adapt to the vicissitudes of the other's behavior and the larger network. As a result, agreements should not only be more frequent under higher than lower mutual dependence, but such exchange also should draw actors attention to their capacity to produce joint benefits and joint control in the negotiation context. Thus, their relation becomes an object for actors, setting up the conditions for ritual behavior symbolic and expressive of a person-to-relation tie. The hypotheses posit a three step causal chain, indicating that more mutual dependence generates more frequent exchange and, by increasing the frequency of exchange over time, more ritual gifts, as follows:

Hypothesis 1:   Given equal power potential (dependence), if the mutual dependence of actors is high rather than low, they engage in more frequent exchange and increase the frequency across time.

Hypothesis 2:   In repetitive negotiations, higher mutual dependence produces more gift giving *indirectly* by increasing the frequency of exchange over time.

Previous research on dyadic negotiations has found that exchange relations with greater mutual dependence tend to foster more conciliatory bargaining and a greater probability of agreement in one-shot negotiations (Bacharach and Lawler 1981; Lawler and Bacharach 1987; Lawler 1992a). In this prior work, each actor had complete information about each other's alternatives. The present study provides actors information only on their own alternative, making it possible to test whether the total-power effect in previous work is contingent on actors' perceiving the dependence of each. These more limited information conditions match the information conditions of Emerson, Cook, and associates (Cook and Emerson 1978, 1984; Cook, Emerson, Gillmore, and Yamagishi 1983); so if mutual dependence affects the frequency of exchange in this study, it indicates a purely structural effect for total power.

Uncertainty Hypotheses

One of the most common explanations for ritual behavior in social constructionist writings is uncertainty reduction. Berger and Luckmann (1966) suggest objectification as a process by which repetitive behavior reduces uncertainty about the future and they portray ritual as an affirmation of the resulting sense of order and regularity. Relatedly, Wuthnow (1987) explicitly views uncertainty in the outside environment as shaping ritual behavior. He offers the following proposition: *Ritual behavior is more frequent and stronger when members of a society or group face uncertainty because ritual evokes a shared, taken-for-granted reality that deals with the uncertainty.* Shared realities embody actors' moral obligations to one another and ritual behavior ostensibly dramatizes and brings forth these obligations in concrete social settings (Wuthnow 1987).

The conditions of uncertainty that promote ritual behavior, according to Wuthnow and Berger and Luckmann, are fortuitously the same sort of conditions that lead to commitment in exchange theory (Cook and Emerson 1984). The focus of the former is the objectification-to-ritual sequence and the focus of the latter is the exchange behavior-to-exchange relation sequence. Exchange theory can explain the development of repetitive exchange among the same actors (i.e., the formation of exchange relations), but it does not explain how their relations then become social objects for actors and the consequence this has for phenomena such as ritual behavior. Lawler and Yoon (1993) theorize that positive emotions produced by frequent exchange account for objectification. This idea produces a basis for understanding uncertainty. Uncertainty should foster objectification because frequent exchange in the context of uncertainty should produce more positive emotion than otherwise.[5]

Applied to a focal dyad within a negatively-connected exchange network, alternative partners in the larger network are relevant environmental conditions. Uncertainty is reflected in part by the subjective probabilities of various profits from exchanges with an alternative. Given constant expected value, the "flatter" the distribution of probabilities across a range of possible exchanges with the alternative, the greater the uncertainty. In this context, Wuthnow's (1978) theorizing and related ideas from exchange theory (Emerson 1981) suggest the following hypotheses:

Hypothesis 3: When actors in an exchange relation are highly uncertain about the benefits they are likely to receive from their respective alternative relations, they engage in more frequent exchange with each other and increase the frequency across time.

Hypothesis 4: In repetitive negotiations, uncertainty about payoffs from alternative relations increase gift giving *indirectly* by affecting the frequency of exchange over time.

Some support for the impact of uncertainty on repetitive dyadic exchange is provided in a laboratory study by Cook and Emerson (1984). They asked whether a network with more alternatives of equivalent value for each actor would produce more frequent negotiated agreements within select dyads. They compared two networks—a four-actor closed circle in which each had two potential exchange partners with an expanded network (12 actors) in which each of the four focal actors had two additional alternatives. The results support the hypothesis that the network with more uncertainty (4 alternatives each and a total of 12 actors) produced more frequent exchange within certain dyads.

To conclude, we suggest that token gifts are most likely when actors in dyadic exchange have substantial power over each other—that is, when the relation has more "total power" or mutual dependence (Emerson 1972; Bacharach and Lawler 1981). Under such conditions, actors are particularly likely to negotiate exchanges yielding each significantly more benefit than available elsewhere; and more frequent exchange, in turn, should increase their tendency to provide each other token gifts "without strings attached" as an expression of an emerging interpersonal relation. We predict the same effects and intervening processes for uncertainty as we do for mutual dependence, though there is a firmer theoretical and empirical backdrop for the hypotheses on power-dependence.

METHOD

This experiment investigates a focal dyad in a negatively-connected network in which each focal actor has one alternative partner, and exchange in the focal relation is likely to be more profitable than exchange with the alternative. The actors have equal power dependence but know only their own dependence, that is, their own alternative. The exchange process involves explicit negotiations (Cook and Emerson 1978; Bacharach and Lawler 1981) rather than nonnegotiated reciprocal transactions (Emerson 1981; Molm 1990). Several, repeated negotiation episodes occur over time in the focal dyad, and while repeated negotiations must occur, repeated agreements need not. The experiment adapts a standard two-party explicit bargaining context (Pruitt 1981; Lawler and Bacharach 1987; Lawler and Yoon 1993).

Experimental Design and Subjects

A 2 x 2 design manipulated mutual dependence (low or high) and uncertainty of payoffs from the alternative partner (low or high). Fifty-two dyads (all females) were randomly assigned to one of the four experimental conditions (13 dyads per cell). Dyads were composed of "real" subjects who bargained with each other across a series of eight independent negotiation episodes. The primary behavioral dependent variables were the frequency of agreement during the first half (first four episodes), the frequency of agreement for the second half (last four episodes), and the number of gifts given during the last half. A post experimental questionnaire measured attitudinal commitment (propensity to stay in the relation), self-reports of positive feelings (pleasure/ happiness and interest/excitement), and the perceived closeness of the exchange relation.

Procedures

Upon arrival, subjects took a seat in separate rooms and read written instructions explaining that they would bargain anonymously with a person in the next room. One of them represented an organization called Alpha attempting to buy a raw material (iron ore), the other an organization called Beta attempting to sell the raw material. Thus, it was an intergroup setting with one issue, the price of iron ore.

Instructions indicated that the study would simulate up to 12 years of negotiation, one negotiation episode per year. The negotiations were separate and independent, because the price set for one year had no formal bearing on that in subsequent years. Each year the two organizations negotiated anew over the price. If an agreement was not reached in a given year, the price paid (by Alpha) or received (by Beta) was determined by an agreement with an alternative supplier or buyer.

As the instructions explained, the two organizations had engaged in preliminary discussions and their offers were quite far apart. Alpha's representative had offered a price of one cent per unit, while Beta's representative had asked for 17 cents per unit. In light of this gap, the subjects' task was to negotiate on behalf of their group's interests. Alpha was to negotiate for as low a price as possible, Beta for as high a price as possible. Subjects' pay depended on the agreement price.

The instructions contained a "profit list," indicating their own group's profits at 17 potential agreement prices, represented by the numbers 1 to 17. Subjects had information only on their own profit stated in terms of points. For Alpha an inverse, linear relationship existed between profit and price levels on each issue, for Beta a positive, linear relationship. Consistent with related work (Lawler and Bacharach 1987; Lawler, Ford, and Blegen 1988), subjects did not have exact information on their negotiation partner's profit at each price level.

The bargaining took place via written offers across a maximum of five bargaining rounds in each year (episode). A round consisted of one offer by each bargainer. When making an offer, subjects had three options: (1) stick with and repeat their last offer, (2) accept the last offer made by the other, or (3) make a counteroffer (i.e., concession). Subjects had to confine offers to one of the 17 price levels and they could not retract earlier concessions.

Negotiation continued until a price agreement emerged on the issue or until the end of the fifth round. If agreement was not reached, subjects received zero points (profit) from their negotiation, but then reached agreement with a simulated other through a drawing (see Bacharach and Lawler 1981; Lawler and Bacharach 1987 for similar procedures).

Gift Option

The initial instructions indicated they would have the option of giving gifts later in negotiations.[6] The later instructions, read after year four indicated "gifts allow you a way to express how you feel about your relationship to [the other]." The instructions likened the gifts to giving a person candy, flowers, or a card acknowledging a relationship. Gifts were made by completing a form at the end of each bargaining episode from years five through eight. On the form, subjects indicated whether they wished to provide a gift or not. The form served as a voucher that subjects could exchange for pieces of candy after the experiment. If subjects did not send a gift, they kept the voucher for themselves and could exchange it for additional pieces of candy. Importantly, the instructions indicated they would not know if the other gave them a gift until the experiment was over. This removed the possibility of subjects treating gifts as an explicit exchange. At the end of the experimental session, the experimenter brought subjects a large container with a variety of candy, and they chose one piece for each gift voucher.

Experimental Manipulations

Low vs. High Mutual Dependence

Mutual dependence was manipulated by varying the expected value of the alternative (see Bacharach and Lawler 1981; Lawler and Bacharach 1987 for similar procedures). The expected value of agreement with the alternative partner was 275 points under high mutual dependence but 375 points under low mutual dependence. Each party had knowledge of only their own alternative. The expected value of the optimal (midpoint) agreement in the focal dyad was 400 points. That they could gain more with the focal than the alternative partner is called for by the scope conditions for our argument. This actually provides a conservative test of mutual dependence effects. Even if alternatives are not as good as the focal relation, we argue that the quality of alternatives provide a varying power potential with implications for gift giving.

Table 1. Manipulations of Mutual Dependence and Uncertainty*

| Points from | High Mutual Dependence | | Low Mutual Dependence | | Points from |
| | Uncertainty | | Uncertainty | | |
alternative	Low	High	Low	High	Alternative
350 or more	1%	14%	1%	14%	350 or more
325	1%	14%	1%	14%	325
300	10%	15%	10%	15%	300
275	76%	16%	76%	16%	275
250	10%	15%	10%	15%	250
225	1%	14%	1%	14%	225
200 or less	1%	14%	1%	14%	200 or less

Note: The midpoint agreement in the focal dyad was worth 400 points to each subject.

Low vs. High Uncertainty

The probability distribution of the payoffs from the alternative manipulated uncertainty. In the low uncertainty condition, the probability distribution was highly peaked (leptokurtic) at the expected value, whereas under high uncertainty, it was flatter (platykurtic). The probability distributions for each experimental condition are in Table 1.

Dependent Measures

Three dyad-level variables reflect the causal steps in our theorizing: frequency of agreement during the first half (agreement frequency-1), frequency of agreement of during the last half (agreement frequency-2), and the number of gifts (summed across actors) during the last half.

The *frequency of agreement* was measured as the frequency of episodes (years) in which focal negotiations yielded agreement during sessions one to four and five to eight. We take the association of earlier (agreement frequency-1) and later frequencies of agreement (agreement frequency-2) as reflecting habitualization.

Gift giving was measured as the average number of gift slips transferred between parties between episodes five to eight. Each actor could give zero or one gift at the end of each year; so, the dyad measure ranges from zero to eight.

Post-questionnaire

Self-report measures of positive emotion or mood were included on the post-questionnaire administered after episode (year) eight. At the time, the subjects completed the questionnaire, they did not know the experiment was over. Subjects reported their feelings along a series of bipolar adjectives; and factor analysis yielded two dimensions that correspond to the pleasure/happiness (Isen 1987; Kemper 1978) and interest/excitement (Izard 1977; Deci 1975). This result is consistent with our earlier study (Lawler and Yoon 1993). The pleasure/happiness index summed four items: pleased-displeased, happy-unhappy, confident-insecure, and contented-discontented (Cronbach's alpha = .86); the interest/excitement summed three items: interested-dull, energetic-tired, and motivated-unmotivated (Cronbach's alpha = .74). Dyad scores were the sum of individual scores.[7]

Other measures include attitudinal commitment and the perceived closeness of the relation (5 items). To measure the attitudinal commitment, we asked two questions: (1) "If you had another chance to bargain over the same issue and you could choose who you bargain with, how likely would you be to choose your present negotiation partner?" and (2) "If you needed to work with someone on a cooperative task and could choose your partner, how likely would you be to choose the person with whom you are negotiating?" The index of attitudinal commitment summed responses to each question and dyad scores summed individual scores. The zero-order correlation between the two items is .62. Items composing an index for perceived closeness of the relation included friendly- unfriendly, cooperative-competitive, close-distant, coming together-coming apart, and team oriented-self oriented (Cronbach's alpha = .87).[8]

RESULTS

Analyses of variance are used to test the predicted main effects on the frequency of exchange. These are the only direct effects predicted by the hypotheses. The mediating role of the frequency of exchange is evaluated with ordinary least-squares regression, testing the indirect effects of mutual dependence and uncertainty on gift giving through the frequency of exchange.

Analyses of Variance

Exchange Frequency

Table 2 contains the mean frequency of agreement by experimental condition for the first four episodes (agreement frequency-1), the last four episodes (agreement frequency-2), and across all episodes (agreement frequency-T). A 2 x 2 analysis of variance for each measure of exchange frequency reveals a consistent pattern of effects—significant main effects for mutual dependence [F's $(1, 51)$ = 10.27, 6.35, and 11.73 and p's < .001, .05, .01 respectively], and

Table 2. Agreement and Gift Frequencies by Experimental Condition (N =52)

	Mutual Dependence			
	High Dependence		Low Dependence	
	Uncertainty		Uncertainty	
	High	Low	High	Low
Agreement Frequency-1	2.23	3.15	1.92	1.54
Agreement Frequency-2	2.85	3.38	2.69	2.08
Agreement Frequency-T	5.08	6.54	4.62	3.62
Gift Frequency	5.54	5.85	5.08	5.08

Note: Agreement frequency-1 indicates it is for the first 4 bargaining episodes (years), agreement frequency-2 for the 5th to 8th bargaining episodes, and agreement frequency-T for all (8) bargaining epiosodes.

dependence by uncertainty interactions [F's (1, 51) = 4.75, 3.96, 6.21, all p's < .05]. The effect for power is consistent with hypothesis 1—the frequency of agreement is greater when actors are more dependent on each other (both have poorer alternatives), and this pattern occurs across both uncertainty conditions despite the interaction. In contrast, the predicted main effect (hypothesis 3) for uncertainty does not occur (all F's < 1).

The pattern of the uncertainty by power interactions reveals the predicted positive effect of uncertainty on exchange frequency under low mutual dependence—higher uncertainty produces more frequent agreements—but the opposite effect occurs under higher mutual dependence. A t-test indicates that the difference in overall exchange frequency between high and low uncertainty conditions is marginally significant under low mutual dependence (t (52) = 1.47, p < .10, one-tailed), while the reverse pattern under high mutual dependence does not reach statistical significance (t = 1.01, ns). The interaction is too tenuous to dwell on, but the trends suggest the plausibility of our uncertainty prediction under lower mutual dependence and, importantly, the pattern for mutual dependence remains the same across both uncertainty conditions. The remainder of the results section will focus on the effects of mutual dependence.

Exchange Frequency by Round Blocks

As an initial indicator of the "habitualization" of exchange, we conducted an analysis of variance by round blocks, the first four episodes constituting one block and the last four the second. The results reveal a block main effect, F (1, 103) = 11.15, p < .01, consistent with the theory. The rate of agreement is greater in the second block than in the first (M = .55 for the first block, .69 for the second block). There are no significant interactions of block with mutual dependence or uncertainty.

Table 3. Standardized Regression Coefficients (OLS)
for Agreement Frequencies and Gift Giving (N =51)

Independent Variables	Dependent Variable		
	(1) Agreement Frequency-1	(2) Agreement Frequency-2	(3) Gift Giving
High Mutual Dependence	.674*** (.106)	.307 (.108)	.057 (.739)
High Uncertainty	.160 (.106)	.210 (.096)	-.115 (.647)
Interaction	-.473* (.150)	-.252 (.141)	0.59 (.949)
Agreement Frequency-1		.421** (.129)	-.243 (.939)
Agreement Frequency-2			.554*** (.971)
R^2	.248	.310	.247

Notes: See Note 1, Table 1.
High mutual dependence, high uncertainty are dummy variables; omitted categories are low mutual dependence and low uncertainty
* $p < .05$, ** $p < .01$, *** $p < .001$; standard errors in parentheses.

Regression Analysis

The primary test of the theory is provided by a regression analysis that determines whether repetitive exchange is a mediating process through which mutual dependence affects gift giving. In this analysis, we test a simple causal chain: High mutual dependence increases the exchange frequency early on (agreement frequency-1) which, in turn, increases later exchange (agreement frequency-2) which, in turn, enhances gift giving. Recall that we conceptualize the path from power to early exchange to later exchange frequency as involving habitualization, and the path from later exchange frequency to gift giving as involving typification and objectification. These are Berger and Luckmann's (1966) main conditions for ritual behavior.

Table 3 contains the results. Consistent with hypothesis 2, the results indicate that mutual dependence (total power) affects gift giving through the growth of repetitive exchange across time. First, the impact of mutual dependence on early exchange frequency is quite strong ($\beta = .674$) and the impact of early exchange frequency on later exchange frequency is also quite strong ($\beta = .421$).

Table 4. Standardized Regression Colefficients (OLS) for Positive Emotions, Closeness of Relation, Attitudinal Commitment (N = 52).

	Dependent Variable			
	(1)	(2)	(3)	(4)
Independent Varables	Pleasure-Happiness	Interest-Excitement	Closeness of Relation	Attitudinal Commitment
High Mutual Dependence	-.066 (.301)	-.158 (.165)	-.184 (.248)	-.092 (.681)
High Uncertainty	-.145 (.263)	-.070 (.145)	-.110 (.248)	-.065 (.560)
Interaction	-.159 (.386)	.215 (.213)	-.114 (.363)	-.099 (.875)
Agreement Frequency-1	.048 (.382)	.152 (.210)	-.218 (.359)	.049 (.865)
Agreement Frequency-2	.421** (.395)	.031 (.218)	.541** (.372)	.465** (.895)
R^2	.262	.030	.269	.246

Notes: See Note 1, Table 1.
High mutual dependence, high uncertainty are dummy variables; omitted categories are low mutual dependence and low uncertainty
* $p < .05$, ** $p < .01$, *** $p < .001$; standard errors in parentheses.

Second, the direct effect of mutual dependence on later exchange frequency is not statistically significant when early exchange frequency is controlled. The impact of mutual dependence on later exchange frequency is indirect, and mediated by early exchange frequency, yielding an indirect effect of .284 (.674 × .421). Third, controlling for antecedent variables (model 3), only later exchange frequency has a significant affect on gift giving. Adding this effect, the indirect effect of mutual dependence on gift giving is .157. The overall implication is support for our main hypotheses about the indirect impact of mutual power dependence on gift giving.

Questionnaire Data

Parallel Regressions

The post questionnaire data on attitudinal commitment, self-reported emotion, and perceived closeness of the relation can be used to corroborate the rationale underlying our hypothesis. There are reasonable zero-order correlations between gift giving and pleasure/happiness (r = .45), perceived closeness (r = .48), and attitudinal commitment (r = .21), making it at least plausible that these are part of a common process. If the same *indirect* process that produces gift giving also produces positive emotion, attitudinal commitment, and perceived closeness, this would add significant weight to the rationale for the hypotheses and to our inferences about gift giving in the study. Thus, we conducted parallel regression analyses that simply substituted attitudinal commitment, positive emotions, and perceived closeness of the relation for gift giving. Table 4 contains these results.

Mutual dependence produces attitudinal commitment (i.e., inclination to stay in the relation) in exactly the same way that it produces gift giving. Controlling for all antecedent variables, only later exchange frequency enhances attitudinal commitment to the relation ($\beta = .465$, p $<.01$). Combining these results with those in Table 3, greater mutual dependence leads to more attitudinal commitment to the relation by increasing early exchange frequency and, indirectly, later exchange frequency. The pattern results for pleasure/happiness is identical ($\beta = .421$, p $< .01$) as are those for perceived closeness of the relation ($\beta = .541$, p $< .01$).

Thus, parallel causal models for gift giving, attitudinal commitment, pleasure/happiness, and perceived closeness of a relation cohere nicely and this is striking. The findings, as a whole, are consistent with our theorizing about expressiveness being produced by frequent exchange and about the role mutual dependence plays as an exogenous, structural condition.

Motivation for Giving Gift

Subjects also were asked on the post-questionnaire to indicate whether they gave gifts because they "felt positive toward the other," because they expected the other "would give gifts to [them]," or to "increase [their] own profits." Of these items, only the first (positive feelings) was correlated significantly with gift giving (r = .41, p $<.01$). The reciprocity and profit motivations had negative but nonsignificant associations with gift giving (r's = $-.20$, $-.15$, respectively). This offers validation to our interpretation of gift giving as expressive rather than instrumental behavior in this study (see Lawler and Yoon 1993 for additional evidence).

DISCUSSION

The theory and research suggests how instrumental and expressive features of dyads can become intertwined and produce affectively-based commitments in exchange networks (Lawler and Yoon 1993). We assume a situation in which people make choices jointly with others, in this case two-party negotiations, and in the process of making joint choices, they experience an "emotional buzz." This emotion is felt individually but it also makes them more aware of something they share—a relation. The relation therefore becomes an object of attachment, setting up the conditions for expressive behavior. As a result, even people making rational choices on the basis of self-interest may become more willing to act on behalf of something larger—a relation, group, or organization. Ritual behavior, such as token unilateral gifts, is a manifestation of this process.

An experimental laboratory setting created a purely token form of gift giving that involved a value domain distinct from the negotiated exchange (see Emerson 1981, 1987). The structural context for examining gift giving was a minimal exchange network (negatively-connected) with two focal actors, who negotiated repeatedly and who each had an alternative partner from whom they could get profitable agreements, though not as profitable as in the focal negotiation. Their degree of dependence on each other (mutual dependence) was either low or high in the context of an equal power dependence relation.

We use Berger and Luckmann's (1966) social-constructionist explanation for ritual behavior to analyze gift giving as a special type of commitment behavior in negotiated social exchange (see Lawler and Yoon 1993). Three theoretical steps are involved in our effort to incorporate social-constructionist ideas within a social exchange framework: (1) conceptualizing repetitive exchange as "habitualizing behavior," (2) theorizing that "objectification of an exchange relation" involves "typification," that is, a definition of the relation in positive terms, and (3) interpreting token gifts as ritual behavior expressive of such an objectified relation. We argue that greater mutual dependence (or total power) in an equal-power relation is especially likely to foster habitualization and typification and, thereby, produce more gift giving in negotiated exchange.

The results of the experiment were consistent with our theoretical hypotheses that mutual dependence should enhance repetitive agreement directly and gift giving indirectly. The first step examined the direct effect of power dependence on the frequency of exchange early in the negotiations. Again, the results clearly and strongly support the power dependence hypothesis—high mutual dependence increased the frequency of exchange early in the repetitive negotiations. The second step was to test the prediction that the frequency of agreement increases from early to later in the negotiations. Again, the results support the prediction. In fact, when later-agreement frequency was regressed on early-agreement frequency, controlling for the manipulated variables, only the early-agreement frequency had a significant effect on the later frequency.

These results are consistent with the notion that mutual dependence fosters the "habitualization" of exchange.

The final and most important step the test of the theory examined was whether gift giving was produced by repetitive exchange. We argue that as the frequency of exchange cumulates or grows, actors "objectify" or "typify" the relation and this, in turn, leads them to initiate more gift giving. Our behavioral and questionnaire results support this theoretical reasoning. First, the regression analysis indicated that controlling for antecedent variables, the frequency of agreement in the later half of the negotiations significantly affected the rate of gift giving; in fact, none of the other variables (mutual dependence, uncertainty, early-agreement frequency) had significant direct effects on gift giving when later-agreement frequency was controlled. Second, a variety of questionnaire measures—attitudinal commitment, self reports of pleasure/happiness, the closeness of the relation with the other—are not only correlated in the expected manner with gift giving, but also are produced through the same causal process as that producing gift giving. We cannot test several aspects of the theoretical rationale for the main hypotheses because the questionnaire was administered at the end of the experimental session, but the results of the questionnaire data are consistent with key parts of the reasoning underlying the behavioral hypotheses.[9]

The results did not support hypotheses on the uncertainty. The hypothesis was that greater uncertainty attached to payoffs from alternative relations would increase the frequency of exchange in the focal relation and, in turn, gift giving. While there was no clear support for this hypothesis, the results did reveal an interesting, albeit weak, pattern. Uncertainty produces more agreements under low mutual dependence, while the pattern is reversed under high mutual dependence, though not statistically significant. This weak interaction might be interpreted as a framing effect (Kahneman and Tversky 1979).

Framing effects stem from the fact that people respond differently to prospective gains and prospective losses. When making a choice between prospective gains, people prefer a sure gain—that is, they avoid risk; thus, when choosing between losses, they prefer risk and choose the option with the less certain losses. These framing effects are diminished if there are large differences of expected value (Kahneman and Tversky 1979). In our experiment, actors were only choosing among prospective gains. Lower mutual dependence should produce the framing effects of prospective gains; thus, when people are less mutually dependent they should be more averse to risk and reach more agreements in the focal relation. Under high mutual dependence the framing effects should be weaker or "wash out," because of a large difference of expected value between the focal and alternative relation. The significant interaction of uncertainty and mutual dependence is generally consistent with this framing interpretation. Frequent exchange among the same actors may involve uncertainty-avoidance behavior instead of uncertainty-reduction behavior, the basis of our original hypothesis.

In combination with an earlier study (Lawler and Yoon 1993), there is now significant support for the idea that structural power (equal vs. unequal dependence and mutual dependence) affects token gift giving indirectly through the frequency of exchange. Lawler and Yoon (1993) provide support for the hypothesis that an emotional/affective process accounts for the positive impact of exchange frequency on gift behavior, and the present study incorporates a social-constructionist account of how the "objectification of the exchange relation," a cognitive process, promotes ritualized gift giving. While the results are consistent with this overarching theorizing, evidence for the objectification process and for ritualization is inferential because we have not directly tested the role of these. Future work should bring together the two dimensions of structural power in a single theoretical formulation and, in this context, explicate further and directly test the predicted relationship between the emotional/affective and objectification processes.

Broader Implications

This theory and research should be viewed as having a complementary relationship to rational choice explanations for commitment formation. The most basic idea from rational choice is that repetitive exchange between the same actors will emerge and continue as long as the expected value of payoffs within the focal relation exceed those available from alternative relations (Elster 1986). The experiment establishes power dependence conditions that produce such a difference in expected value, and one can interpret the impact of mutual dependence on repetitive exchange in rational choice terms. Our theory suggests, however, that in the course of producing repetitive exchange, power dependence (structural) conditions also engender incipient group formation in the dyad due to the emotional/affective consequences of actors jointly dealing with and resolving negotiation problems.

Using rational choice principles, social exchange theory can provide a good explanation for how and why a pattern of repetitive exchange comes about, but not for the emotional and affective consequences of repeated and profitable exchange. Moreover, neither exchange theory nor other related perspectives, such as transaction-costs economics, can explain actors' tendency to give each other token benefits "without strings attached." The explanation in our theory is that the objectification of the exchange relation leads to behavior expressive of that relation. The process of objectification starts with the "emotional buzz" involved in actors accomplishing a joint task, this makes the relation more salient as an object and a target for affective attachment (Lawler 1992b; Lawler and Yoon 1993; Markovsky and Lawler 1994). The basic result is emergent group formation within the dyad or "incipient institutionalization" in Berger and Luckmann's (1966) terms. Incipient institutionalization promotes the initiation of ritual behavior symbolic of the relevant social entity, in this case, the dyad.

Overall, the research suggests how endogenous processes in dyads within a minimal exchange network, produce a particularly important form of commitment behavior: gift giving. For theoretical reasons, we focus on the early stages of an exchange relation and treat the initiation of gifts where actors do not know if the other is similarly inclined. The purpose is to understand how and when exchange relations *begin* to take on expressive properties. When this occurs, the emerging tie within the dyad is likely to have ramifications for the larger network. If members of the dyad interact more with each other and become affectively attached to their relation, then their relations with others in the larger network will change, and the relation of those others to still others also will change. Endogenous processes, like those we identify, are a starting point for important micro-to-macro effects in exchange networks.

In a recent theory of group solidarity, Markovsky and Lawler (1994) indicate that when an actor experiences positive emotion in repeated interaction with members of a group, they will begin to view the group as an object and become attached to it as well. The objectification of an exchange relation may spread in this way to a larger group or network and serve as a foundation for group-oriented action (see also Lawler 1992b). Our theory predicts that objectification of the larger group would enhance the per capita rate of gift giving among group members and, more generally, what Organ (1990) and others have termed "organizational citizenship behavior," that is, the willingness to do extra, unrequired, and uncompensated things.

An important further question is how "incipient institutionalization" in an exchange relation might give rise to more "sedimented institutionalization." Berger and Luckmann (1966, pp. 57-60) suggest that habitualization and typification are sufficient for incipient institutionalization but that a "third force" is necessary for emerging institutional patterns to "harden." The third force may constitute a specific or generalized other or a referential structure that justifies and legitimizes the developing patterns of behavior. At the level of a dyad, "sedimented institutionalization" is implied if the dyadic relation becomes so close that members not only reduce contact with others in the network in favor of interaction in the focal relation but act as if these alternatives are not present or are irrelevant. One indicator of such a condition is a decrease in the degree that actors attend to or think about the alternative during negotiations.

The post-questionnaire of the experiment contains some relevant evidence. One item asked actors how much they thought about the alternative during the negotiations and another asked whether the alternative became more or less important to them in the later episodes of bargaining. The results are generally consistent with the behavioral data. Higher mutual dependence reduced the degree that subjects reported thinking about the alternative and (F = 3.95, p < .05) and under low uncertainty in particular, higher mutual dependence reduced the perceived importance in the later episodes (interaction effect F = 4.21, p < .05). Also as one would expect, gift giving was negatively associated with each of these items (r's = −.27 and −.26, p's < .05 and .06, respectively). Thus, there is some indication that incipient group formation promoted the sort of perceptions important to sedimentation or the "hardening" of institutionalization in Berger and Luckmann's (1966) terms.

Another condition, reflecting a "hardening of institutionalization," is that the relation exerts a moral/normative constraint on actors. The transition from incipient to sedimented institutionalization is a possible way to examine how and when the emotional/affective processes produce a relation that exerts a moral/normative constraint on actors. In our theorizing, relations that "enable" actors to jointly resolve problems and produce mutual benefits ostensibly become objects for positive feelings of accomplishment (see Lawler 1992b). Projecting such a process further, we hypothesize that exchange relations which "enable" actors to do things (i.e., provide opportunities for choice, etc.) also come to "constrain" them as moral obligations are associated with the relation and informal or formal sanctioning emerges. The moral/normative character of exchange relations can develop from emotional/affective processes that, in turn, have their source in rational choices about who to exchange with. This is a broader view of how micro social orders develop from negotiated exchange. Future theoretical and empirical work should address this.

To conclude, exchange networks create differential power-dependence among actors or positions and also varying levels of mutual dependence among component dyadic relations. The paper suggests, theoretically and empirically, how mutual dependencies in a social structure can foster incipient commitment and gift giving. The underlying process is that frequent exchange among the same actors engenders positive emotion and leads them to objectify the dyadic relation and engage in rudimentary forms of ritual behavior. Thus, structurally-based dependencies, by shaping frequencies of exchange in given dyads, stimulate emotional and cognitive processes that add expressive components to instrumental exchange relations.

ACKNOWLEDGMENT

This study was funded by the Duane C. Spriestersbach Professorship while held by the first author at the University of Iowa. The order of authorship between Baker and Large was determined by the flip of a coin. We thank Jodi O'Brien and Barry Markovsky for helpful comments. Address correspondence to Edward J. Lawler, Department of Organizational Behavior, New York State School of Industrial and Labor Relations, Cornell University, Ithaca NY 14853.

NOTES

1. We assume a dyad embedded in a larger exchange network and focus solely on exchange that is explicit and negotiated (Cook and Emerson 1978; Lawler and Yoon 1993) rather than implicit and nonnegotiated (Molm 1990, 1992). The exchange opportunity occurs repeatedly, given the social structure (i.e., network), and endogenous processes within dyads are the proximal causes of ritual behavior.

2. Gift giving is central to what Ekeh (1974) terms generalized exchange, exemplified by the Kula Ring (Malinowski 1922). The contrast of gift exchange with negotiated exchange in Emerson (1981) and Akerlof (1982) resembles Ekeh's (1974) contrast between generalized and restricted exchange. The idea is that gifts both reflect and reproduce trust, commitment, and cohesion in instrumental relations (e.g., Akerlof 1982, 1984; Ekeh 1974). We take this to imply that if people are willing to make token, unilateral gifts, an element of expressiveness has been introduced into a purely instrumental exchange relation.

3. If third parties legitimize such a reciprocal typification over time, the institutional objects "harden and thicken," that is, become sedimented (see Berger and Luckmann 1966, p. 59). This should enhance the constraining effect of a relation or group in an emerging micro order.

4. The proposition assumes a key idea from Lawler and Yoon (1993)—namely that repetitive exchange fosters objectification through an emotional process. People ostensibly get an "emotional buzz" from accomplishing a joint task with others, such as reaching agreements, and this makes their relation more salient as a unit. This process was documented empirically in a prior study (Lawler and Yoon 1993); here we assume it and focus on the objectification-to-ritual link.

5. The implications of uncertainty have also been examined in organizational studies of contracting (Williamson 1975, 1981; Pfeffer, Salanczik, and Leblebici 1978; Ouchi 1979). Organizational studies suggest that uncertainties due to the lack of information and the presence of opportunism increase transaction costs and thereby make market price mechanisms inefficient. Given such uncertainties, contracts internalize market transactions into a hierarchical organizational structure (Williamson 1975, 1981) or act as a substitute for trust in transactions (Okun 1981). Whatever the organizational form or underpinnings, the broad arguments in this literature dovetail with our approach. A key difference is that we offer a social-constructionist account that emphasizes the emotional/affective consequences of repeated agreements (contracts) between the same individual actors.

6. These initial instructions prevented later instructions from being a surprise. Lawler and Yoon's (1993) study indicated that this early mention of the gift giving option did not produce a "mental set."

7. Lawler and Yoon (1993) propose pleasure/happiness and interest/excitement as two facets of positive emotion. Based on Izard (1977) and recent analyses of the circumflex model of emotion (Watson 1988; Watson, Clark, and Tellegen 1988; Watson and Tellegen 1985), they define pleasure/

happiness as "feeling gratified" and interest/excitement as "feeling energized." Consistent with these expectations, varimax rotation with Kaiser normalization showed two factors with the following factor loadings: For the pleasure/happiness dimension, pleased-displeased (.94), happy-unhappy (.91), confident-insecure (.56), and contented-discontented (.79); for the interest/excitement dimension, interested-dull (.73), energetic-tired (.78), and motivated-unmotivated (.84). The zero-order correlation between the two dimensions was −.091.

8.   Principal component factor analysis confirmed one factor and the factor loadings are as follows: friendly-unfriendly (.79), cooperative-competitive (.68), close-distant (.76), coming together-coming apart (.86), and team oriented-self-oriented (.82).

9.   In Lawler and Yoon (1993), equal power was compared to unequal power (holding total power or mutual dependence constant), and the results were that equal power produced more interest/excitement indirectly through more frequent exchange; equal power did not increase pleasure/happiness. The current study held equal power constant and found that more mutual dependence in the relation indirectly produced more pleasure/happiness, but there was no effect on interest/excitement. Both studies indicate that positive emotions mediate the impact of structural power, but different dimensions of positive emotion are important. Perhaps, greater mutual dependence (total power) enhances the pleasure/happiness derived from negotiated agreements, whereas inequalities of dependence dampen interest/excitement developed within the negotiation process.

REFERENCES

Akerlof, G. A. 1982. "Labor Contracts as Partial Gift Exchange." *Quarterly Journal of Economics* 97: 543-69.

———. 1984. "Psychological and Sociological Foundations of Economics." *American Economic Review* 74: 79-82.

Arrow, K. J. 1972. "Gift and Exchange." *Philosophy and Public Affairs* 1: 343-62.

Bacharach, S. B. and E. J. Lawler. 1981. *Bargaining: Power, Tactics, and Outcomes.* San Francisco, CA: Jossey-Bass.

Berger, P. L. and T. Luckmann. 1966. *The Social Construction of Reality.* New York: Doubleday.

Blau, P. M. 1964. *Exchange and Power in Social Life.* New York: Free Press.

———. 1977. "A Macrosociological Theory of Social Structure." *American Journal of Sociology* 83: 26-54.

Callero, P. L. 1991. "Toward a Sociology of Cognition." Pp. 43-54 in *The Self-Society Dynamic: Cognition, Emotion, and Action,* edited by J. A. Howard and P. C. Callero. New York: Cambridge.

Caplow, T. 1982. "Christmas Gifts and Kin Networks." *American Sociological Review* 47: 383-92.

———. 1984. "Rule Enforcement without Visible Means: Christmas Gift Giving in Middletown." *American Journal of Sociology* 89: 1307-23.

Cheal, D. J. 1986. "The Social Dimensions of Gift Behavior." *Journal of Social and Personal Relationships* 3: 423-39.

———. 1988. *The Gift Economy.* New York: Routledge.

Collins, R. 1975. *Conflict Sociology.* New York: Academic Press.

———. 1981. "On the Microfoundations of Macrosociology." *American Journal of Sociology* 86: 984-1014.

————— . 1989. "Toward a Neo-Meadian Sociology of Mind." *Symbolic Interaction* 12: 1-32.
Cook, K. S. and R. M. Emerson. 1978. "Power, Equity, and Commitment in Exchange Networks."
    American Sociological Review 27: 31-40.
————— . 1984. "Exchange Networks and the Analysis of Complex Organizations." Pp. 1-30
    in *Research on the Sociology of Organizations* (Volume 3), edited by S. B. Bacharach and
    E. J. Lawler. Greenwich, CT: JAI Press.
Cook, K. S., R. M. Emerson, M. R. Gillmore, and T. Yamagishi. 1983. "The Distribution of
    Power in Exchange Networks." *American Journal of Sociology* 89: 275-305.
Cook, K. S. and J. M. Whitmeyer. 1992. "Two Approaches to Social Structure: Exchange Theory
    and Network Analysis." *Annual Review of Sociology* 18: 109-27.
Deci, E. 1975. *Intrinsic Motivation.* New York: Plenum Press.
Durkheim, E. 1915. *The Elementary Forms of Religious Life.* New York: Free Press.
Ekeh, P. 1974. *Social Exchange Theory.* Cambridge, MA: Harvard University Press.
Elster, J. 1986. *Rational Choice.* New York: New York University Press.
Emerson, R. M. 1972. "Exchange Theory Part II: Exchange Relations and Networks." Pp. 38-
    87 in *Sociological Theories in Progress* (Volume 2), edited by J. Berger, M. Zelditch, Jr.,
    and B. Anderson. Boston, MA: Houghton-Mifflin.
————— . 1981. "Social Exchange Theory." Pp. 30-65 in *Social Psychology: Sociological
    Perspective*, edited by M. Rosenberg and R. H. Turner. New York: Basic Books, Inc.
————— . 1987. "Toward a Theory of Value in Social Exchange." Pp. 11-46 in *Social Exchange
    Theory*, edited by K. S. Cook. Newbury Park, CA: Sage Publications.
Goffman, E. 1967. *Interaction Ritual.* New York: Doubleday.
Gouldner, A. 1960. "The Norm of Reciprocity: A Preliminary Statement." *American Sociological
    Review* 25: 161-78.
Haas, D. F. and F. A. Deseran. 1981. "Trust and Symbolic Exchange." *Social Psychology
    Quarterly* 44: 3-13.
Heath, A. 1976. *Rational Choice and Social Exchange.* Cambridge, England: Cambridge
    University Press.
Homans, G. C. 1961. *Social Behavior: Its Elementary Forms.* New York: Harcourt Brace and
    World.
Isen, A. M. 1987. "Positive Affect, Cognitive Processes, and Social Behavior." Pp. 203-53 in
    Advances in Experimental Social Psychology (Volume 20), edited by L. Berkowitz. New
    York: Academic Press.
Izard, C. E. 1977. *Human Emotions.* New York: Plenum Press.
Jasso, G. 1993."Choice and Emotion in Comparison Theory." *Rationality and Society* 5: 231-
    74.
Kahneman, D. and A. Tversky. 1979. "Prospect Theory: An Analysis of Decision under Risk."
    Econometrica 47: 263-91.
Kemper, T. D. 1978. *A Social Interactional Theory of Emotions.* New York: Wiley.
Kollock, P. and J. O'Brien. 1992. "The Social Construction of Exchange." Pp. 89-112 in *Advances
    in Group Processes* (Volume 9), edited by E. J. Lawler, B. Markovsky, C. Ridgeway, and
    H. Walker. Greenwich, CT: JAI Press.
Kramer, R. M. 1991. "Intergroup Relations and Organizational Dilemmas: The Role of
    Categorization Processes." *Research in Organizational Behavior* 13: 191-228.
Lawler, E. J. 1992a. "Power Processes in Bargaining." *Sociological Quarterly* 33: 17-34.
————— . 1992b. "Choice Processes and Affective Attachments to Nested Groups: A Theoretical
    Analysis." *American Sociological Review* 57: 327-39.
Lawler, E. J. and S. B. Bacharach. 1987. "Comparison of Dependence and Punitive Forms of
    Power." *Social Forces* 66: 446-62.

Lawler, E. J. and R. Ford. 1993. "Bargaining and Influence in Conflict." In *Sociological Perspectives on Social Psychology*, edited by K. S. Cook, G. A. Fine, and J. S. House.

Lawler, E. J., R. Ford, and M. A. Blegen. 1988. "Coercive Capability in Conflict: A Test of Bilateral Deterrence vs. Conflict Spiral Theory." *Social Psychology Quarterly* 51: 93-107.

Lawler, E. J. and J. Yoon. 1993. "Power and the Emergence of Commitment Behavior in Negotiated Exchange" *American Sociological Review* 58: 456-81.

Levi-Strauss, C. 1969. *The Elementary Structures of Kinship*. Boston: Beacon Press.

Malinowski, B. 1922. *Argonauts of the Western Pacific*. London: Routedge and Kegan Paul.

Markovsky, B. and E. J. Lawler. 1994. "A Theory of G roup Solidarity." In *Advances in Group Processes* (Volume 11), edited by B. Markovsky, K. Heimer, and J. O'Brien. Greenwich, CT: JAI Press.

Mauss, M. 1954. *The Gift: Forms and Functions of Exchange in Archaic Societies*. New York: Free Press.

Molm, L. 1987. "Extending Power Dependence Theory: Power Processes and Negative Outcomes." Pp. 179-98 in *Advances in Group Processes*, edited by E. J. Lawler and B. Markovsky. Greenwich, CT: JAI Press.

————. 1990. "Structure, Action, and Outcomes: The Dynamics of Power in Social Exchange." *American Sociological Review* 55: 427-47.

————. 1992. "Dependence and Risk: Transforming the Structure of Social Exchange." A Paper Presented at the Annual Meeting of American Sociological Association, Pittsburgh, August.

Okun, A. M. 1981. *Prices and Quantities: A Macro-Economic Analysis*. Washington, DC: Brookings Institute.

Organ, D. W. 1990. "The Motivational Basis of Organizational Citizenship Behavior." *Research in Organizational Behavior* 12: 43-72.

Ouchi, W. G. 1979. "A Conceptual Framework for the Design of Organizational Control Mechanisms. *Management Science* 25: 833-48.

Pfeffer, J., G. R. Salancik, and J. Leblebici. 1978. "Uncertainty and Social Influence in Organizational Decision Making." Pp. 306-332 in *Environments and Organizations*, edited by M. W. Meyer. San Francisco: Jossey-Bass.

Pruitt, D. G. 1981. *Negotiation Behavior*. New York: Academic Press.

Rabbie, J. M., J. C. Schot, and L. Visser. 1989. "Social Identity Theory: A Conceptual and Empirical Critique from the Perspective of a Behavioral Interaction Model." *European Journal of Social Psychology* 19: 171-202.

Schwartz, B. 1967. "The Social Psychology of the Gift." *American Journal of Sociology* 73: 1-11.

Thibaut, J. C. and H. H. Kelley. 1959. *The Social Psychology of Groups*. New York: John Wiley and Sons, Inc.

Titmuss, R. 1971. *The Gift Relationship*. New York: Pantheon

Turner, V. W. 1982. *Celebration: Studies in Festivity and Ritual*. Washington, DC: Smithsonian Institute.

Watson, D. 1988. "Intraindividual and Interindividual Analyses of Positive and Negative Affect: Their Relation to Health Complaints, Perceived Stress, and Daily Activities." *Journal of Personality and Social Psychology* 54: 1020-30.

Watson, D., L. A. Clark, and A. Tellegen. 1988. "Development and Validation of Brief Measures of Positive and Negative Affect: The PANAS Scale." *Journal of Personality and Social Psychology* 54: 1063-70.

Watson, D. and A. Tellegen. 1985. "Toward a Consensual Structure of Mood." *Psychological Bulletin* 98: 219-235.

Williamson, O. E. 1975. *Markets and Hierarchies: Analysis and Antitrust Implications: A Study in the Economics of Internal Organization.* New York: Free Press.

———. 1981. "The Economics of Organization: The Transaction Cost Approach." *American Journal of Sociology* 87: 549-577.

Wuthnow, R. 1987. *Meaning and Moral Order: Explorations in Cultural Analysis.* Berkeley, CA.: University of California Press.

Research in the Sociology of Organizations

Edited by **Samuel B. Bacharach**, *New York State School of Industrial and Labor Relations, Cornell University*

Associate Editors: **Peter Bamberger**, *Bar Ilan University,* **Pamela Tolbert**, *Cornell University* and **David Torres**, *University of Illinois, Chicago Circle*

REVIEW: "Research in the Sociology of Organizations is designed as an annual review of current work related to organizations. The editor apparently identifies individuals with major scholarly programs related to organizations and invites them to summarize their work. All six chapters in this, the second volume, clearly reflect several years of work by the authors. All are intellectually independent; other than emphasizing something about organizations, they have little in common.

-- *Contemporary Sociology*

Volume 12, Special Issue on Labor Relations and Unions
1993, 309 pp. $73.25
ISBN 1-55938-736-X

Edited by **Ronald Seeber**, *Cornell University* and **David Walsh**, *Miami University*

CONTENTS: Introduction, *Samuel Bacharach, Ronald Seeber, and David Walsh.* Issues in Union Structure, *George Strauss.* Integrating U.S. Labor Leadership: Union Democracy and the Ascent of Ethnic and Racial Minorities and Women into National Union Office, *Daniel B. Cornfield.* The Formation and Trasformation of National Unions: A Generative Approach to the Evolution of Labor Organizations, *Peter D. Sherer and Huseyin Leblebici.* National Union Effectiveness, *Jack Fiorito, Paul Jarley, and John Thomas Delaney.* Unions and Legitimacy: A Conceptual Refinement, *Gary N. Chaison, Barbara Bigelow, and Edward Ottensmeyer.* Conflict Resolution and Management in Contemporary Work Organizations; Theoretical Perspectives and Empirical Evidence, *David Lewin.* A Diagnostic Approach to Labor Relations in Organizations, *Arie Shirom.* The Labor Movement as an Interorganizational Nework, *David J. Walsh.* Associational Movements and Employment Rights: An Emerging Paradigm?, *Charles Heckscher and David Palmer.*

Also Available:
Volumes 1-11 (1982-1993) $73.25 each

Advances in Group Processes

Edited by **Edward J. Lawler**, *Department of Organizational Behavior, Cornell University*

REVIEWS: "A major impression one gets from this volume is that far from being dormant, the social psychology of groups and interpersonal relations is quite vibrant, and very much involved with compelling problems.

"Concerns about the imminent demise of group processes as an area of study in social psychology are clearly exaggerated. But should doubts remain, they ought to be allayed by the range and quality of the offerings in this volume."

" . . . should be of interest both to specialists in group processes and to sociologists who are interested in theory, particularly in theoretical linkages between micro and macro analysis. Because many of the papers offer thorough reviews and analyses of existing theoretical work, as well as new theoretical ideas, they are also useful readings for graduate students.

-- *Contemporary Sociology*

Volume 11, 1994, 239 pp. $73.25
ISBN 1-55938-857-9

Edited by **Barry Markovsky, Karen Heimer,** and **Jodi OBrien,** *Department of Sociology, University of Iowa.*

Also Available:
Volumes 1-10 (1984-1993) $73.25 each

Advances in group processes